C0-AUT-018

Why I Am Not a Scientist

THE PUBLISHER GRATEFULLY ACKNOWLEDGES THE
GENEROUS SUPPORT OF THE GENERAL ENDOWMENT FUND
OF THE UNIVERSITY OF CALIFORNIA PRESS FOUNDATION

why I am **NOT** a scientist

Anthropology and Modern Knowledge

Jonathan Marks

UNIVERSITY OF CALIFORNIA PRESS
Berkeley Los Angeles London

University of California Press, one of the most distinguished university presses in the United States, enriches lives around the world by advancing scholarship in the humanities, social sciences, and natural sciences. Its activities are supported by the UC Press Foundation and by philanthropic contributions from individuals and institutions. For more information, visit www.ucpress.edu.

University of California Press
Berkeley and Los Angeles, California

University of California Press, Ltd.
London, England

Library of Congress Cataloging-in-Publication Data

Marks, Jonathan (Jonathan M.), 1955–
Why I am not a scientist : anthropology and modern knowledge / Jonathan Marks.
p. cm.
Includes bibliographical references and index.
ISBN 978-0-520-25959-1 (cloth : alk. paper) — ISBN 978-0-520-25960-7 (pbk. : alk. paper)
1. Anthropology—Philosophy. 2. Science—Philosophy. 3. Evolution (Biology)—Philosophy. I. Title.

GN33.M32 2009
301.01—dc22 2008040508

Manufactured in the United States of America

18 17 16 15 14 13 12 11 10 09
10 9 8 7 6 5 4 3 2 1

This book is printed on Natures Book, which contains 30% post-consumer waste and meets the minimum requirements of ANSI/NISO Z39.48–1992 (R 1997) (*Permanence of Paper*).

For Peta and Abby

Contents

Preface

All right, I admit it—I *am* a scientist. I can no more define myself out of that category than Bertrand Russell could define himself out of Christianity, in his 1927 essay to which the title of this book is an homage—"Why I am not a Christian."

Lord Russell tried to define himself out of Christianity on narrow grounds, maintaining his otherness on the basis of denying the existence of God and immortality and maintaining a fair degree of ambivalence about Jesus himself.

On the other hand, he wasn't thinking culturally.

He was, after all, the product of Christian society, Christian history, Christian morality, Christian literature, and—Jesus Friggin' Christ!—even Christian language. His national anthem was "God Save the King." He matriculated at Trinity College. Bertrand Russell wasn't a Muslim, a Jew, or a Hindu. He wasn't a Trobriand Islander or a Khoe bushman. Whatever he found repugnant or offensive in aspects of Christianity, he was inescapably in a global sense a Christian, albeit one with some doctrinal issues.

That is the sense in which I am a scientist. I treat the natural and supernatural realms as disconnected from one another. I don't think much about God, and I certainly wouldn't want to share those thoughts with you. However, I do capitalize His (or Her) occupation, by convention.

On the other hand, I have some beefs with science—and I consider them significant enough to have some reservations about fully identifying with it—as Russell had with Christianity. These include the idea that all other knowledge, and all other forms of knowledge production, are illegitimate; that large classes of people perhaps ought to be political and social inferiors because they are natural inferiors; and that any critical analysis of any aspect of science reveals an anti-science agenda. The first is ethnocentrism, the second is racism or sexism, and the third is paranoia. They should be no more welcomed in science than they would be at a city council meeting or a family reunion.

Why, after all these years, is there still scientific racism? Why do scientists raise the same classes of data today that they did decades ago, and which were illegitimate then? Why are they taken seriously when they do so? And if they can't be taken seriously about this, then why should they be taken seriously about anything else?

And that's only the tip of the iceberg.

Why is it still so threatening to learn that you can't believe everything a scientist says? We knew that long ago. And yet to talk about it today is commonly perceived as producing not so much a deeper understanding of science as rather a threat to science—and that is a shame. The real question is, Given that there are circumstances under which proclamations of scientists are unreliable, what are those circumstances? In other words, if science is a system of thought and action, and the descriptive and comparative analysis of different systems of thought and action is anthropology's stock in trade, then why shouldn't the development of a relativizing anthropology of science be seen as an advancement for the science of anthropology, and indeed for our understanding of science in general?

Science is widely accepted to be three different things: a method of understanding and of establishing facts about the universe; the facts themselves, the products of that method; and a voice of authority and consequently a locus of cultural power. This triple identity creates tensions within science and conflicting roles for it.

This book explores science as a set of beliefs and practices about nature and knowledge that developed in the seventeenth and eighteenth centuries in Europe—the most significant being that the subject of science, nature, can and ought to be approached independently of what we might now call supernature (i.e., the realm of spirit and miracle rather than of matter and law) and culture (i.e., the realm of human contrivance—law, beauty, morals, politics, and the like).

The biggest obstacle to studying science anthropologically is the choice of whether to universalize it or to particularize it. Did the Classic Maya have science? Is science something that everybody has in their fashion but only certain peoples exercise strongly? Or is science something that only "we" have? In which case, what do "they" have?

I think the most reasonable approach is to acknowledge that everyone has knowledge about the world, much of it accurate, which allows them to manipulate their environments in diverse and productive ways. Science, however, is a particular approach to knowledge that is more precisely localized in the cultural history of Europe. I don't hold it against anybody that they don't have "science"; it is simply that, if we extend the label beyond the traditions of thought that developed in early modern Europe, then we still have to come up with a word to encode the distinction between Western more or less modern science (on the one hand) and Inuit, Maya, medieval Islamic, or Nuer science (on the other). The distinction is real; the question then is, What do you call "our" system of knowledge production? I side with those who would call it "science," in contrast to "other non-science systems of knowledge."

Laura Nader discusses the directions anthropology has taken in engaging with science, where British anthropologists were initially more active than their American counterparts.[1] One direction (since the 1950s) lay in documenting the considerable knowledge about their environment possessed by indigenous peoples; the other (since the 1980s), in ethnographic studies of scientists as subjects. A third, more recent direction involves the engagement of local people specifically with science. She omits, however, that anthropology—or at least large segments of it—is self-consciously "science." Somehow, unfortunately, those segments have rarely if ever connected with the "anthropology of science."[2] Genetics within anthropology, for example, tends to be represented as biological anthropologists aspiring to be "real" geneticists and as cultural anthropologists studying "real" geneticists.[3] The biological anthropologists engaged in genetic research and yet also engaged with anthropological (as opposed to biological) issues are a small bunch.[4] Paleoanthropology has contact with indigenous rights, alternative narratives of origins, and a history rich in colonialism and racism—but little engagement with them anthropologically.[5] Primatology has had the greatest anthropological engagement with what it does.[6] Yet within all of these anthropological sub-subfields, there also exists a contemporary reactionary literature that is so reductive and so unreflexive as to be almost embarrassing to have classified as anthropology—and yet still claims the authority of science.

I have taught biological anthropology—the basic course on human evolution—at four universities, but students got science credit for the course at only two. That bothers me—not because I care about being formally labeled as a scientist, but because it has implications for understanding, and for communicating, what is constituted by science.

How can you teach students about science if you're not sure whether you actually do it? Imagine the creationists taking the line that they are not out to subvert science education after all, because at the local university the courses on human evolution don't actually count as science!

This book is not intended as a comprehensive review of all the work that has proceeded in the anthropology of science.[7] It is, rather, intended as a necessarily somewhat idiosyncratic synthesis, drawing on themes I have been interested in as a biological anthropologist and as a general anthropologist. There isn't a canon, and I am simply trying to draw a bit from the diverse literatures in science studies, sociology of science, history of science, cultural studies, and of course science (or whatever biological anthropology is).

This book is most fundamentally about the relationship between anthropology and science. It tries in some measure to reconcile the two, by fitting anthropological science into an anthropological frame. If it succeeds, the final product will tell us not only a bit more about anthropology but a bit more about science as well.

Finally, an advocate of scientism—that is to say, the largely uncritical acceptance of everything said with the authority of science—might also be called a "scientist." I'm definitely not one of those. What I want to know is, When can't you believe everything a scientist says?

Parts of this book were written during my stay as a visiting fellow in the Genomics and Society Forum at the University of Edinburgh. For comments on the whole manuscript I thank Peta Katz, Jim Bindon, Laurie Nelson, Deborah Bolnick, and Stefan Helmreich. Several people looked at bits of this book, and I thank them for their comments: Hugh Gusterson, Karen Strier, Ian Tattersall, Jonathan Kahn, Susan Lindee, and Bob Sussman.

Charlotte, North Carolina
July 2008

ONE Science as a Culture and as a "Side"

Many years ago, in the late 1980s, as a postdoc in genetics at the University of California at Davis, I was interviewed by National Public Radio on the subject of the Human Genome Project, then beseeching Congress and the American public for a few billion dollars.

Sure, it would keep molecular biologists employed into the foreseeable future, but was it science?

Of course not, I told NPR, with the assuredness that comes with having recently earned a doctorate and of working in a laboratory with radioactive isotopes, toxic chemicals, and expensive machines with flashing multicolored lights. Science involves *testing hypotheses;* we all know that. We teach our students that. The Human Genome Project wasn't testing any hypotheses—it was merely collecting a large mass of data because we now could. We had the capability to carry out a big molecular genetics project, but it was disconnected from science; it was not the way we were taught that science was supposed to be.

I was also speaking with the scientific reasonableness that came from knowing we had recently beaten down a challenge to science by creationists, who had been working to get their ideas accorded equal time alongside evolution in science classes. A few years before the Human Genome Project discussion, a federal judge named William R. Overton had ruled that creationism wasn't science and therefore should not get equal time—or any time at all—alongside Darwinism. Why? He had been told by respected scientific authorities that science tests hypotheses and creationism does not.

A few years later, I was recruited to help review a few hundred grant proposals by the scientific society Sigma Xi, which gives small sums to graduate students starting their thesis research. I was instructed to divide the proposals into two piles: those that tested hypotheses, and those that did not. The ones in the first pile would get about five hundred dollars each, and the ones in the second pile would not.

So, if you did not test a hypothesis, you could be denied five hundred dollars, but you could get three billion. And a creationist who did not test a hypothesis was *not* doing science, but a molecular biologist who did not test a hypothesis was *indeed* doing science.

This made little sense to me at the time, and it makes little sense to me now. I have no doubt that the Human Genome Project is (or was) science, and little doubt that creationism is not science. But testing hypotheses does not seem to have much to do with it. So what does? What makes something science? What makes something nonscience, like humanities—which are respectable and scholarly but nevertheless differentiated from science? And what makes something pseudoscience—that is, something disreputable?

A DEFINITION OF SCIENCE

Let us begin with a definition: *Science is the production of convincing knowledge in modern society.*

This is what I mean when I use the word *science,* and all of the polysyllabic words in that definition merit some discussion. By using *production* we acknowledge that science is not a passive experience. Scientific knowledge is a product—and as a product it is the result of some process. That process is science, and it is what we mean to analyze. There is a subtler and more threatening point embedded in this recognition, however. If science is the active production of something—say, reliable information about the universe—then it is more than, or at least different from, mere discovery. Discovery is a passive operation: to a suitably primed observer, the fact merely reveals itself.

But of course, "facts" of nature, of the universe, cannot reveal themselves, for they cannot act. The act of discovery hinges on what is constituted by a scientist's being "suitably primed." Being ready for a discovery implies a context of the right social environment, the means, and the intellectual precursors that allow the discovery to be rendered sensible. It is unlikely that natural selection could have been discoverable outside the context of competitive, industrial Victorian England. At any rate, it had never been discovered before and was recognized separately by Charles Darwin, Alfred Russel Wallace, and Herbert Spencer at about the same time. The fertilizing union of egg and sperm could not be discovered until the invention of microscopes. In the absence of the germ theory of disease, the initial serendipitous discovery of antibiotics would have passed silently.

Such examples illustrate that the production of scientific knowledge is highly context-specific, and that it is the context, more than the particulars of the discovery, that are critical. The individual discovery (or discoverer) is not terribly important, for if Darwin had never lived we would still have natural selection; if Watson and Crick had never lived we would still have the structure of DNA (the great chemist Linus Pauling was only weeks away from figuring it out himself).

If you can discover something only when you are ideologically, technologically, and intellectually prepared for it, then it seems to follow that the interesting question for understanding science is not "How was the fact discovered?" but rather, "What was needed in order to recognize and identify the fact?" Since facts are now seen to be actively produced rather than passively revealed, the production of facts becomes something that we can study, as one would study any other social or cultural process.

By *convincing,* we mean that there is a social process beyond mere discovery or fact production. Somehow the fact has to be accepted, in order for other scientists to incorporate it and build upon it. While it is certainly true that the growth and progress of science are due in part to the community at large recognizing that somebody's work is "correct," it is also true that some ideas we now know to be correct have sometimes been slow to be accepted (such as continental drift), and ideas that we now know to be wrong have sometimes been rapidly and widely accepted (such as Piltdown man).

These mistakes would not exist if science proceeded simply by the rejection of wrong ideas and their supplanting by right ones. Moreover, the mistakes can be rendered invisible by the pretense that science actually works that way—that is, merely figuring out what's true—which serves to conceal the networks of communication, authority, and power that retard or augment the spread of knowledge. By focusing on science as specifically convincing knowledge, we call attention to the processes that render its facts visible and credible to others. Revelation, for example, is a real source of knowledge, but not of convincing knowledge, for the knowledge can be shared only by someone who has had a similar revelatory experience. Science is different in that its work is directed toward the goal of successfully convincing an open-minded outsider of its propositions.

By *knowledge,* we mean reliable information about the universe. It is something you can bank on. Of course, it could be wrong. But if it were wrong too frequently or too egregiously, it wouldn't be very reliable. So science is information about the universe that comes with some source of authority behind it. The authority is different from a shaman's, or from the pope's, or from a policeman's, and consequently its source merits some reflection.

Finally, by *modern society,* we mean the ideas, values, and social practices that arose in Europe and its satellites and colonies at a time in the eighteenth century often referred to as the Enlightenment. As the name suggests, we look back on this time as an era of illumination, when formerly obscure things finally "came to light." There are, of course, reasons why such conceptual changes occurred at that time—it is not as if there were simply more geniuses being born—but, more significant, it was there and then that science as we recognize it today began to take form. This is not to denigrate the thought and work of people from other times and places but simply to note that what we now call science is not directly descended from their thought and work but rather from the work and thought of those proverbial "dead white males."

We may certainly admire the metallurgy of ancient West Africans, the astronomy of ancient Mesoamericans, or the architecture and philosophy of the ancient Greeks or Chinese, but none of these achievements represented science in any easily recognizable form. The cultural differences among these peoples, especially in relation to the Euro-Americans of the Enlightenment, are far larger than the superficial similarities that emerge from the fact that they all thought deeply about the natural world and applied the results of that effort successfully in diverse ways.

Science is different, and began to emerge only with a strange idea of the Enlightenment: that the physical world—the world of perceptions and sensations and measurability—was somehow different and separate from the spiritual and moral worlds. Nature was amenable to certain forms of knowledge production of a different order than the kinds of knowledge one could obtain from the spiritual realm. This was not to say that God or heaven did not exist, only that they were separate and distinct from the physical world. This bracketing off of nature from supernature became the signature of science. One was subject to measurable forces and deep regularities; the other was capricious, miraculous, and unknowable. Or, at least, knowable in a very different way.

One example can serve here: The ancient Greeks, for all of their contributions to our knowledge, had no word for religion. It is not that they were not religious, or that they lacked confidence in divine spirits and beings, but rather that these forces permeated their lives so inextricably that it made no sense to bracket them off from the mundane, earthly aspects of their lives and worlds.

In fact, the division of nature from supernature, of the physical universe from the metaphysical one, has been unfamiliar to most people over most of the course of human history.

Biblical Hebrew had a word, *ruach,* translated into Greek as *pneuma* and subsequently into English as *breath* or *wind.* (You can easily see the association between *breath* and *pneumonia.*) The same word is, however, also translated into English as *spirit* from both the ancient Hebrew and ancient Greek. Of course, *breath* and *spirit* are associated as well, but what mostly seems to divide them is the invisible barrier between the physical world of breath and the metaphysical world of spirit. Without such a barrier as a part of one's conception of the universe, spirit and breath might well be the same thing.

It is specifically the construction of that invisible barrier which differentiates what we recognize as science from other kinds of thinking about the world and manipulating it, even those of our own more remote cultural ancestors. Thus, we restrict *science* to mean specifically the kind of thinking that arose in Europe in the seventeenth and eighteenth centuries, when the respective domains of nature and supernature began to be circumscribed, in contrast to the more widespread view of seeing them as mutually interpenetrating and porous—indeed, as not really different from one another.[1]

THINKING ABOUT SCIENCE

People are always up in arms about science education and science literacy in ways that they don't seem to be up in arms about humanities education and humanities literacy.[2] C. P. Snow was a distinguished physicist at Cambridge as well as a successful novelist (his wife, Pamela Hansford Johnson, was an even better novelist). Ever since his essay "The Two Cultures" appeared in the 1950s,[3] the academy has been forced to acknowledge that the price we pay for knowing more and more about the universe is that knowledge becomes so specialized that a scientist often knows nothing *but* science.

Finding himself astride two distinctly different, although both highly intellectual, social circles, Snow set out the proposition that academic life was increasingly becoming bifurcated. On the one hand, the humanists on campus were becoming increasingly distant from, and uninterested in, the latest developments in science (which included the new areas of computers, space flight, and molecular biology). On the other, scientists were becoming increasingly removed from art, literature, and aesthetics, the very things that make us "human."

In short, Snow said, these two groups of scholars were at the point of becoming distinct campus cultures, a term he self-consciously chose to make the analogy to anthropology clear. They think about the world differently, have different interests, languages, value systems, and can hardly communicate meaningfully with one another. What Snow saw at Cambridge University was a microcosm of what he believed was going on in society generally. There were otherwise smart people who knew about quantum physics but had once rather pathetically "tried a bit of Dickens"—as if Dickens were particularly deep—and there were also smart people who could not even articulate the Second Law of Thermodynamics (that physical systems tend toward entropy or disorder, or leak energy, which is why a perpetual motion machine is impossible), a situation Snow equated with having never read anything by Shakespeare.

Like other classic texts, Snow's essay has been read differently by different audiences. In one self-interested reading, Snow is seen to be railing against the ignorance of science by humanists.[4] Perhaps the oddest thing about that particular reading is that it undermines Snow's title and very theme—the two cultures—by which he clearly intended to convey the relativistic notion of difference without hierarchical ranking; this reading replaces Snow's insight with its conceited, ethnocentric opposite.

A leading British literary critic lashed out at Snow for presuming that any sort of equivalence was appropriate between a mere equation and the works of Shakespeare, which express the grandest and basest of human motivations and articulate the deepest and most resonant feelings we all share. And, he added for good measure, Snow wasn't even all that good a novelist.[5]

In recent years, discussions about the science-humanities divide in academics and in life generally have built on Snow's essay, and it is difficult to find anyone defending the idea that the situation is improving. On campus, humanists publish books, scientists publish articles. As a result, humanists tend to read books, and scientists tend to read articles. Sure, there is the occasional scientist who has read a novel in the past year (science fiction, of course, doesn't count) and the occasional humanist who happily slogs through the latest issue of *Scientific American* to keep abreast of superstring theory and what's new in short interspersed elements of the genome—but they are, relatively and absolutely, a small minority. The rest of us find it more than a full-time activity to keep up to date on our subspecialty (say, molecular anthropology), much less on our specialty (biological anthropology), much less on our general field (anthropology, or whatever it actually says on the diploma on the wall), still less on other sciences—and still have the time and mental energy to read novels.

This "cultural" difference is manifested in other ways as well. In the late 1950s, the National Science Foundation was still relatively new, and cold war anxieties, aggravated by the Soviet orbiting satellite, Sputnik, were promoting both the expansion of science and the massive transfer of financial support for it from the shoulders of private foundations to the government bureaucracy. Not surprisingly, the academy began to change, as did its priorities. Major universities that had long been known as centers for training and research began to rely on research grants as a major source of their operating budget. This in turn placed a new emphasis on the ability of new faculty to apply successfully for grants to fund their research. Indeed, anything else they did began to be downplayed; a science department's significance became measurable in terms of the amount of grant money it brought in. And if the quality of teaching, or even the structure of the curriculum, suffered thereby, it was just an unfortunate consequence of the ruthlessly competitive marketplace, in which scientific grants dictated both academic stature and clout within the university.

I didn't realize it at the time, but that was why, as an undergraduate at a prestigious research university, I had a professor who could barely make himself understood in English trying to teach me integral calculus; a professor who couldn't explain anything sensibly trying to teach me introductory biology; and a professor who never looked at the class trying to teach me cell biology.

It wasn't at all the same in the humanities. In the humanities there was no money—or, at least, very little. Consequently, your advancement wasn't predicated quite so much on getting grant moneys. If you did, that was a feather in your cap, but it wasn't like biochemistry, where you were simply expected to. In both the sciences and the humanities you were expected to publish, but publication in the sciences required significantly more investment of capital. Consequently, faculty in the sciences were being recruited more and more one-dimensionally, as departments loaded up on faculty in hot (i.e., fundable) areas of research. If the curriculum suffered, so be it.

By the time I began teaching at Yale in the 1980s, the biology department had nobody who could teach undergraduate anatomy. Anatomy had to be taught in the anthropology department. Weirder still, students who took, say, Comparative Primate Anatomy did not get science credit for it. They got social science credit, since the home department, anthropology, was officially a social science department. It didn't matter that the course was, by anyone's definition, science; that it was a laboratory class; that it was competently taught; and that it was filling a significant gap in the curriculum. What mattered was that *we needed to patrol the boundaries of science—particularly in its bureaucratized form—and aggressively regulate it.* Even if the regulation was arbitrary and produced bizarre results, the ability to decide what counted as science was a form of social power that was not to be surrendered lightly.

I was even more confused when a popular course I taught on the evolution of human behavior was not considered science, although human evolution was my specialty, while a similar course offered in the biology department, by someone whose specialty was *not* human evolution, *was* considered science. What did that imply about science? That it was anything an officially designated scientist said, regardless of their expertise?

The answer was made a bit clearer in 1995, when a mathematician and a biologist published *Higher Superstition: The Academic Left and Its Quarrels with Science.* This book represented one front of what came to be known as the "science wars" in the mid-1990s and suggested that there was some sort of anti-science conspiracy on the part of creationists, animal rights activists, philosophers and historians of science, and literary critics—in short, on the part of seemingly anyone who had anything remotely critical to say about any aspect of science. Although by no means a prominent member of it, I considered myself a habitué of the "Academic Left," and yet I was unaware of any such conspiracy. The book's tone was odd, rather more like something you might expect from the Inquisition than from a product of more recent times. But what struck me most was a thought experiment the authors suggested. If the humanities faculty of a university

> were to walk out in a huff, the scientific faculty could, at need and with enough released time, patch together a humanities curriculum, to be taught by scientists themselves. It would have obvious gaps and rough spots to be sure, and it might with some regularity prove inane; but on the whole it would be, we imagine, no worse than operative. What the

opposite situation—a walkout by the scientists—would produce, as the humanities department tried to cope with the demand for science education, we leave to the reader's imagination.

To my imagination, at least, the result would probably be a science curriculum with "gaps and rough spots" as well, but one on the whole far better *taught* than previously. One of the consequences of basing decisions about hiring and promotions for science faculty strictly on funding and research, after all, as is customary at major universities, is the widespread devaluation of the quality of teaching in the sciences. Scientists generally receive far less experience teaching while in graduate school, and fewer teaching responsibilities while on the faculty, than do their counterparts in the humanities. If practice makes perfect, then an average scientist might be expected to develop into a less perfect educator than an average humanist. Moreover, if academic scientists are at all as smart as they're cracked up to be, then they certainly realize that their professional fate rests with funding and research, and consequently any time they spend improving themselves pedagogically would act against their own professional interests.

So, on what basis could one realistically expect that scientists would teach a humanities curriculum more competently than humanists would teach a science curriculum? There is only one basis on which to expect that, namely, if sciences are simply hard and humanities are easy. Scientists would necessarily then be smarter than people in other fields and could be expected to pick up those fields more readily than a nonscientist could pick up science.

Maybe that's true. I don't know. But it's not the kind of thing I'd publicly crow about, because it sounds kind of egotistical, arrogant, and boorish. And like a popular unflattering stereotype of scientists.

THE SOKAL HOAX

Okay, this is hilarious. I hope you're sitting down.

There's this journal called *Social Text,* which publishes a lot of this left-wing humanistic stuff. It's a scholarly opinion journal. This scientist at New York University, a physicist named Alan Sokal, decides to show what dopes these humanists are, so he sends them a manuscript called "Transgressing the boundaries: Toward a hermeneutics of quantum gravity," which is full of double-talk and bullshit. The editors are delighted that a physicist apparently wishes to contribute something to the dialogue about science that they are trying to create.

So anyway, they suggest some revisions but figure that since the point of the journal is not the dissemination of new science and since the author is a physicist at a reputable university, he probably knows what he's talking about when it comes to physics, and so they publish it.[6] And as soon as the article comes out, he goes public with the story that he got a totally bullshit article published in their journal.[7] Ha, ha, ha!

Man, that was great! Did those guys look stupid! Score one for our side!

The "Sokal hoax," as it came to be known during the summer of 1996, showed science in a very strange light, scoring points at the expense of its university colleagues. Not only were there two cultures, but one had seemingly declared war on the other. Journalists didn't have to look hard to find scientists who could scarcely contain their glee about a paper that made humanists look so foolish.[8]

But it actually sounded rather more like the final revenge of that antisocial geek with the plastic shirt-pocket protector (another popular unflattering stereotype of scientists). After all, what kind of person goes out of his way to show how smart he is by humiliating others? And then gloats over it? It was that combination of malice and arrogance that left a bad taste in people's mouths.

The paper, obviously, had been submitted under false pretenses. It is an assumption of the scholarly process that one is dealing with a scholarly submission in good faith. Once the good faith agreement is violated, history shows quite clearly that it's not all that hard to get a bogus paper published in the scientific literature. So the Sokal hoax shows nothing about whether the standards are lower in sciences or humanities, or whether one or the other is easier to fool.

But who ever heard of an art historian trying to make biochemists look foolish? Why would they bother?

And yet the distinguished physicist Steven Weinberg could write, "Like many other scientists, I was amused when I heard about the prank."[9] Apparently even mature and subdued scientists were amused before, or instead of, being appalled. Thus Sokal's act, which would ordinarily be regarded as sociopathic, was actually resonant with (at least major parts of) the scientific community.

How had we gotten to such a point, from merely the wry observations of C.P. Snow a few decades earlier? What was the source of such open hostility between the two cultures?

Sociologist Dorothy Nelkin found several factors at work, all related to the erosion of an informal contract between science and modern society.[10] One was a widespread public call for greater accountability on the part of scientists, in place of the honor system that had long been the norm. A second was the large infusion of financial support from "private sector" interests, with attendant claims upon a scientist's loyalty, in turn affecting the public's perception of scientists as a source of unbiased knowledge. Another was the overall relativizing of scientists in society. At least in the last of these, the role of humanists can be discerned, with ethnographic techniques broadly adopted to study science and scientists, just as one would study the origin and production of knowledge in any other culture.

Perhaps this rise of the "anthropology of science" entails a bit of iconoclasm, if we try to study scientists as we study the Yanomamo of Brazil or the Hopi of Arizona. Some scientists find it insulting or degrading (which should make you wonder how the Hopi feel about it, at the very least). Some find it valueless, as if there were something an outsider might see about human behavior that would be invisible to an insider (which, obviously, is a major rationale for more than a century of serious ethnographic research on anybody). Some find what humanists write about science to be impenetrable and jargon-laden. Alan Sokal himself complained that their "incomprehensibility becomes a virtue; allusions, metaphors, and puns substitute for evidence and logic." And, finally, there seems to be a widespread insecurity that humanists have proved, or have convinced themselves, that there is no external reality—that all is perception, or text, or politics.

Sokal, for example, invites "anyone who believes that the laws of physics are mere social conventions" to step out of the window of his twenty-first-floor apartment. The like-minded Oxford biologist Richard Dawkins likewise baits his self-designated academic antagonists to go their scholarly meetings on flying carpets rather than on airplanes.

While it is widely appreciated that there is a social or cultural construction of reality, that does not mean that there are no laws of nature, or that there is "nothing out there." It *does* mean that it may be difficult to distinguish facts from meanings, and that facts are, at the very least, expressed through the medium of language, which is how humans most fundamentally impart meanings to things.

Notice that Sokal did not even say "social construction" but "mere social convention"—which clearly implies something very different—an entirely arbitrary agreement to say "God bless you" after someone sneezes, for example. Once again, that is not at all what we mean by "culturally constructed." We mean that scientific facts are produced, and exist, within a historical and social matrix of meaning. Thus, while no humanist would deny that falling twenty-one stories out of a building is likely to be fatal, our understanding of falling is a cultural contingency. When the earth was thought to be the center of the universe, falling down meant being drawn to the center of the universe. In the seventeenth century, after Jupiter was shown by Galileo to have its own moons, a question became visible that had previously been concealed: if you fell down on Jupiter, would you fall back toward earth or back toward Jupiter?

The point is that nobody had to "discover" falling; it was always there. But what we *think* about falling, that is to say, the science of falling, is derived in large measure from Isaac Newton's construction of it. And Newton's work could only succeed the work of Copernicus and Galileo, for their own ideas made Newton's possible and indeed continue to render Newton's work meaningful. To the extent that more recent work has superseded or generalized Newton's, it is nevertheless contingent upon Newton's. (It is hard to imagine the history of physics bypassing Newton, for he is so iconic, but sooner or later someone else discovers everything.)

THE CULTURAL CONSTRUCTION OF KNOWLEDGE: "WE ARE APES"

A nice example of the problem in confronting the constructed nature of scientific facts can be seen in an essay published in the *Journal of Molecular Evolution* in 2000. It is a minimally referenced essay, labeled "Opinion" and written by the editor in chief, the distinguished biochemist Emile Zuckerkandl.[11] That it is an opinion piece in a journal that rarely publishes them, that it contains but a single reference, and that it was submitted by the editor in chief himself are all relevant. It suggests a very important issue—perhaps commenting upon a brilliant new discovery in molecular evolution?

Alas, no. The piece is about "social constructionism" and rails against an article published in a different journal by the paleontologist Stephen Jay Gould. Gould, of course, did not deny that there is a reality, but whatever he did say was sufficient to get Zuckerkandl inflamed to self-publish a response. Zuckerkandl begins by superciliously drawing distinctions between "the process of discovery" and the early and late stages of maturity and stability of scientific knowledge. "Society does intervene in some important ways in the acquisition of scientific knowledge," he concedes, "yet, at the end of the day, none of these ways affects the content of the scientific product." Since these claims are unreferenced, we cannot know whether he believes this is common knowledge or is simply making it up, oblivious to the difficulties in making such broad declarations.

As luck would have it, I was already engaged in studying a scientific fact that Zuckerkandl himself had discovered (the extraordinary genetic similarity of human and ape) and was busy writing a book on the cultural construction of that natural fact. Zuckerkandl had found that, when you compare the amino acid sequence that constitutes the protein part of hemoglobin—which transports gases in the blood—between human and gorilla, you find only two differences out of 287 possibilities. Thus, he wrote, "from the point of view of hemoglobin structure, it appears that gorilla is just an abnormal human, or man an abnormal gorilla, and the two species form actually one continuous population."[12]

The distinguished paleomammalogist G.G. Simpson responded bluntly. If any competent biologist can tell a human from a gorilla at thirty paces, does it not follow that the "standpoint of hemoglobin"—which seems to confuse the human and the gorilla—is a rather silly standpoint to take?[13] And yet, over the ensuing decades, the "standpoint of hemoglobin"—or relations as told from molecules, as opposed to the animals in which they are found—became so dominant that the biologist Jared Diamond could write a best-selling book calling us the "Third Chimpanzee," predicated on that very genetic comparison.

Actually, though, it is not entirely clear that the discovery of the genetic near-identity of human and ape is strange or paradoxical in the first place. In fact the genetic relationship basically replicates the anatomical relationship: in the great panoply of life's diversity, humans and apes are very, very similar, yet diagnosably different, throughout. The idea that this has a self-evident meaning, which is somehow counterintuitive, is simply the result of two cultural facts: our familiarity with the ape's body, and our unfamiliarity with genetic comparisons.

We have, after all, been studying chimpanzees scientifically since 1699. When they were new and interesting, back in the eighteenth century, scholars ranging from the Swedish biologist Carl Linnaeus to the French social philosopher Jean-Jacques Rousseau and the Scottish jurist Lord Monboddo were overwhelmingly impressed by the striking physical similarity of ape and human. Linnaeus, Rousseau, and Monboddo were all quite satisfied to understand the ape as a variant kind of human—one lacking certain of the essential features of humanity, to be sure, but nearly human nonetheless.

A couple of centuries later, the physical differences between human and ape had become fairly well understood. Anyone who knows what to look for can easily distinguish the femur (thighbone) of an ape from that of a human, although they might look identical to a naïve viewer. The observer might *not*, however, be very inclined to present the relationships of the thighbone as a single number, a scalar quantity. How do you reduce a comparison of three-dimensional forms into a one-dimensional number?[14] Gene sequences, on the other hand, are long chains of simple subunits; their differences are easy to tally and quantify because they are conceptualized in a single dimension, a line. There is charm in comparing linear quantities; everything is either higher than, lower than, or equal to everything else. Witness the popularity of the linearized IQ in the twentieth century, which could be easily compared and rank-ordered, as a stand-in for intelligence, which cannot be so easily compared. So it is with genetic sequences: the extent of their differences can easily be represented numerically, but it is a crude stand-in at best for the overall relationships of the species.

Moreover, it is not clear—as Simpson argued—whether the perceived genetic relationship is transcendent or just erroneous. The fact that biochemically or genetically you might not be able to tell human and ape apart does not necessarily mean that they are identical; rather, it might just mean that the differences between them have not yet been fully studied and evaluated. Indeed, the same pattern is actually present genetically and anatomically: each corresponding part is very similar, yet diagnosably different, in human and ape.

In other words, Zuckerkandl's discovery that human and ape are merely abnormal variants of one another, surprisingly similar from the standpoint of hemoglobin (or protein and DNA sequences more generally), was a highly constructed fact. It is true enough that humans and chimps are more than 98 percent genetically identical, but it is not necessarily true that this is (a) more than, say, the similarity of a human and chimp femur, (b) "realer" than any sort of comparable measurement of the femur, or (c) higher than we should have anticipated.[15]

Another fact can help contextualize the genetic similarity of human and ape: the structure of DNA constrains us to be more than 25 percent genetically identical, in a base-for-base comparison, to a carrot. But saying on that basis that we are genetically "very abnormal carrots" or "over one-quarter carrot" would properly be considered idiotic.

Where, then, is the logic for assuming that the extent of our DNA matching is a measurement of our "true," "deep," or "real" similarity? The DNA matching is an arbitrary measurement, not necessarily highly informative, not obviously highly profound, and rendered meaningful or significant only in a cultural context that privileges genetic information, mystifies genetic information, and privileges scalar comparisons.

Far from being a "lost cause," as Zuckerkandl condescendingly put it, constructionism is what allows us to make sense of his own work; his own un-self-consciousness about it is the shortcoming that prevented him from understanding it himself.

CHANGING TIMES

There does seem to be a time within memory when the science faculty were rather more introspective and less haughty. In 1954, a botany professor published the results of a small informal study: he asked fifteen biologists a set of questions at their Ph.D. orals: Can you identify (1) the Renaissance, (2) the Reformation, (3) the Monroe Doctrine, (4) Voltaire, (5) the Koran, (6) Plato, (7) the Medici family, (8) the Treaty of Versailles, (9) Bismarck, and (10) the Magna Carta. He considered barely one-third of the answers to be satisfactory.[16]

A year later, the geneticist Conway Zirkle went so far as to construct this mock diploma:

> THE JOHNS HOPKINS UNIVERSITY
> certifies that
>
> *John Wentworth Doe*
>
> does *not* know anything but
> Biochemistry
>
> Please pay no attention to any pronouncement he may make on any other subject, particularly when he joins with others of his kind to save the world from something or another.
> However, he has worked hard for this degree and is potentially a most valuable citizen. Please treat him kindly.[17]

Of course, we should be reluctant to generalize from these two examples, but two things are clear. First, they were both published in the journal *Science,* the leading general science periodical in America, which suggests that they had some broad resonance with the scientific community at large. And second, those very graduate students who couldn't identify Plato or Voltaire and knew nothing but biochemistry in the mid-1950s had matured into the tenured gatekeepers of science by the mid-1980s.

What seems to have happened is that the ignorance or benign neglect of other areas of scholarship, noted by C. P. Snow and Conway Zirkle in the 1950s, had metastasized into the paranoid fear and loathing of the "science wars" a few decades later. To be sure, as the old saying goes, just because you're paranoid doesn't mean nobody is out to get you. And indeed there are forces working to undermine aspects of science education—most prominently, creationists; but also (with diverse motivations and credibilities) zealous animal rights activists, greedy corporations, ambitious politicians, sanctimonious anti-abortionists, not to mention just old-fashioned hucksters. But where is the wisdom in imagining that they are all colluding, when they are simply pursuing diverse agendas that happen occasionally to line them up against the perceived best interests of science?

RIVERRUN

You know you're in trouble when the novel's first word is recognizable but unfamiliar and is not even capitalized. It violates the most basic rules of English prose. You are going to have to work hard to get something out of it. Will it be worth the effort?

In this case, the novel is James Joyce's *Finnegans Wake,* and the general consensus is that it is indeed worth the effort.

But why should the onus be on me to have to work so hard to read a story? Stories are supposed to be easy to understand; they're supposed to be *stories.* Not like science, for example, where we take it for granted that years of study are required to master the vocabulary and concepts, where meanings will be hidden from all but the fully initiated.

On the other hand, why shouldn't specialists in things other than science require a specialized vocabulary and conceptual apparatus to communicate their ideas as well?

One of the weirder fronts on the "science wars" is the claim that science strives for transparency while the humanities seem to be striving for opacity, with dense, self-important academic blather, often in the name of "deconstruction" or "postmodernism," rather than the lucid, comprehensible prose that characterizes science. Indeed (this position continues), the very goal of science is to be as widely understood as possible, while these postmodern humanists are terrible writers, merely using gobbledygook to cover up the fact that they have nothing to say.

I suppose that some humanists indeed have nothing to say but need to say something in order to keep their paychecks coming. But that situation is not much different in science.

And just how lucid is the prose in science, anyway? There is certainly very good science writing and very bad science writing. But what is the middle like?

Here's what I think: The writing, on average, is probably better in humanistic fields than in scientific fields. Why? For a simple reason: scholars in humanistic fields have been subject to a lot more intensive formal training in writing than scholars in scientific fields have. It's part of their curriculum.

In a book called *Fashionable Nonsense,* that mathematics wag Alan Sokal returns with a French physicist as coauthor to call attention to the incomprehensibility that characterizes the writing of some humanists and is taken seriously by others.[18] The fact that some of this work is difficult to comprehend (and, to make things worse, some of it is actually translated from French) is, however, a red herring. The real issue is whether there are some useful ideas behind the work.

Once again, some comparative perspective may be useful. Isaac Newton's 1687 *Principia,* which helped to frame modern science, was (and remains) incomprehensible to all but a very few readers. A famous story holds that a Cambridge student, passing Newton on campus, told a friend, "There goes the man that writt a book that neither he nor any body else understands."[19] Nor was that an accident. Newton later remarked that he had made his work unintelligible deliberately "to avoid being baited by little smatterers in mathematicks"[20]—that is to say, by those very people who now man the front lines of the "science wars."

None of this is intended either to demean Newton or to assert some sort of equivalence between the seventeenth-century physicist and the twentieth-century literary critics. All I want to show is that obscure writing is by no means the exclusive domain of contemporary deconstructionists or postmodernists. Needless abstruseness has made a home in science from the beginning.

Nor is there any evidence that the situation is improving. While there are well-known public complaints about the poverty of academic writing in general, there are at least as many bemoaning the poverty of writing specifically in the sciences. As an essay in *Nature* explained, "Pleas for scientists to write readably have failed for at least 300 years." What reason do we have for thinking the future of scientific prose looks any brighter?

> Everyone can write, so it is assumed that writing is easy, or unimportant. Everyone can paint as well, but not everyone's paintings are worth hanging on walls. To expect scientists to produce readable work without any training, and without any reward for success or retribution for failure, is like expecting us to play violins without teachers or to observe speed limits without policemen. Some may do it, but most won't or can't.[21]

In his farewell editorial after a quarter-century of editing the leading science journal in the world, Sir John Maddox was hardly one to mince words, but he could only speculate on the cause of the problem he observed:

> It used to seem that *Nature*'s contributors wrote clearly, but no longer. . . . The obscurity of the literature now is so marked that one can only believe it to be deliberate. Do people hide their meaning from insecurity, for fear of being found out or, in the belief that what they have to say is important, to hide the meaning from other people?[22]

The latter choice, of course, was Newton's—although the journal *Nature* was not yet in existence when Newton lived—but he obviously was a precedent setter. And it's hard to deny the editor of *Nature;* there are no doubt plenty of insecure contributors, and some contributors afraid to be found out, as well.

Sometimes you can even forget delving into the prose itself, for you can't even get past the title. From a random issue of the prestigious *Proceedings of the National Academy of Sciences,* you can easily find a title like "*In vitro* assembly of the undecaprenylpyrophosphate-linked heptasaccharide for prokaryotic N-linked glycosylation," as evocative of James Joyce as you could hope for, albeit certainly unintentionally.

I pulled that one out of the current issue. Another article in the same issue is titled "Giant-block twist grain boundary smectic phases." I know what most of the words mean (except "smectic," which sounds vaguely scatological but actually has something to do with the arrangement of molecules in a liquid crystal). Together, however, they sound like nonsense, with too damn many nouns in a row *(block, twist, grain, boundary).* In combination they sound almost like "Colorless green ideas sleep furiously"—the linguist Noam Chomsky's famous example of a sentence that is recognizably English in spite of being nonsensical.

From the same issue, we can find a stylistic device commonly used in the humanities as well: the two-part title, divided by a colon. In the humanities, this is generally structured as something cute, colon, then something explanatory. In the *PNAS,* however, we can find nothing either cute or explanatory in "Surface-mounted altitudinal molecular rotors in alternating electric field: Single-molecule parametric oscillator molecular dynamics." (I double-checked to see if there is an article missing before "alternating." There isn't.)

And finally, we encounter the newest trend in science titles, the declarative sentence in lieu of the topic: "C-type natriuretic peptide inhibits leukocyte recruitment and platelet-leukocyte interactions via suppression of P-selectin expression." Since you now know the conclusion, the authors seem to be saying, you don't even have to bother reading the article itself; we're sure you have better things to do with your time.

No, far from being a transparent, accessible, universal literary genre, the scientific literature is for the most part as dense and impenetrable as a Mayan codex, and certainly no less so than the humanities literature, postmodern or not. Moreover, it has its own stylistic rules and literary conventions, in some cases so at odds with actual practice that the immunologist Peter Medawar once famously pronounced the scientific paper *as a genre* to be fraudulent (see chapter 4).

TOWARD AN ANTHROPOLOGY OF SCIENCE

Let us adopt the relativistic position that C. P. Snow suggested half a century ago: science is an anthropological "culture"[23] and, by implication, can therefore most profitably be understood using anthropological methods, conceptual frameworks, and analyses.

At the famous Scopes trial (see chapter 5), the attorney for science asked the attorney for religion a set of questions designed to show the latter's ignorance. One question was the number of people alive at the time of Christ. The attorney for religion had never given it any thought. When pressed, he finally said, "When you display my ignorance, could you not give me the facts so I would not be ignorant any longer?" The attorney for science answered him sharply, "You know, some of us might get the facts and still be ignorant."

This gets to the very heart of science. Science is a method, a way to knowledge, a path to enlightenment. Facts are great, but they don't constitute science; they are merely its many endpoints. Science is how we get facts, not the facts themselves. You can know a lot of them yet still be ignorant or unscientific.

This raises a fundamental question about science education. If science is a process of knowledge production, then is science education best expressed as teaching students the process or as teaching them the knowledge itself? If we focus on teaching students the accumulated knowledge, the facts of science, then we are not actually teaching them science. Rather, we are teaching them science's products, and indeed we are misleading them by substituting the teaching of scientific facts, as if it were the teaching of science itself.

Consequently, beware of people who complain about this generation's lack of "science literacy." The kids who don't know the difference between fluorine, chlorine, and schmorine are no worse off than the ones who think Rodin is a Japanese movie monster and that Plato's most lasting contribution is the children's modeling clay which now bears his name. In fact there is probably a large overlap between the kids who are illiterate about science and those who are illiterate about anything else.

All right, there are a lot of people out there who don't know, or don't believe, what you want them to. Is it worse that they don't know whether the Axis won or lost World War II or that they don't know the difference between a muon and a gluon?

Different people know different things. If you don't like what they know, then it stands to reason that the solution is a massive campaign of indoctrination, or evangelization. By that token, though, science will have devolved into an ideology, or a set of beliefs, requiring something like a Nicene Creed to proclaim one's adherence to. The depth of one's knowledge would be a measure of the depth of one's immersion into the faith, and consequently the minutiae of the faith would begin to assume a disproportionate role—thus, the stereotypical science teacher obsessed with the minutiae that Robert Benchley satirized as "The Sex Life of the Polyp," back in the 1920s.

The alternative is not to worry about science literacy, except as an expression of general ignorance. As C.P. Snow originally observed, scientists know their stuff, and humanists know their stuff (to which we might add, the people that know *neither* may well be able to fix the scientist's and the humanist's broken transmission).

Instead, however, let us focus on science as a method of knowledge production. Then learning science is not principally about learning *what* scientists think but *how* scientists think. If science is method, then let us understand how the method works—how it is that science does come to tell us what the physical universe really is like, either because of, or in spite of, its practitioners—and why it is important for us to know what the universe really is like in the first place. There are, after all, other things worth knowing: good from evil, for instance. Legal from illegal. Sublime from vulgar. Gothic from Romanesque.

The point is not that scientists are stupid, which of course most are not; nor that there is not an external reality, which of course there is; nor that science is not the best way of finding it, in which it has achieved considerable success; nor that science is not important, which it manifestly is. The intelligence of scientists, the existence of reality, the methods of assessing that reality, and the importance of doing so do not require defense or justification.

What an anthropology of science raises are more down-to-earth issues. How is scientific knowledge produced? How is science different from other cultural systems that produce knowledge? Can you believe everything a scientist says? If not, why not? How can you tell science from stuff that is not science? How can you tell good science from bad? What constitutes scientific practice—the activities of information gathering, social interaction, and ratiocination—that result in scientific knowledge? What counts as acceptable practice, and why? How does science impinge upon daily life, and how do people adapt to it? How is science absorbed, performed, utilized, and administered in particular political economic contexts? How and why would people resist science? And, from a practical standpoint, are there intellectual areas in which the training of today's scientists could stand some improvement?[24]

TWO The Scientific Revolution

The publication dates of 1543 and 1687 are generally used to bracket the Scientific Revolution. Even if we downplay the intellectual ferment before 1543 (the prior half-century had witnessed the discovery of the Americas and the beginning of the Protestant Reformation, after all) and subsequent to 1687 (the next century would finally bring "Enlightenment"), it is still hard to consider it much of a "revolution," since it took a full 144 years to transpire. That is probably why nobody referred to it as a revolution until the middle of the twentieth century.[1]

Its revolutionary nature is visible principally in hindsight and when played in fast-forward. The two retrospectively significant works published in 1543 were the speculations of a Polish astronomer, Copernicus's *On the Revolutions of the Celestial Spheres,* and the iconoclastic observations of a Belgian anatomist, Vesalius's *On the Fabric of the Human Body.*

Copernicus argued that perhaps the earth-centered Ptolemaic solar system had become a bit too cumbersome, and that perhaps the observations of the heavenly bodies could be at least as easily explained with the sun at the center of the solar system. There are two powerful myths about Copernicus's work: that it was more empirical, persuasive, and accurate than its alternative; and that it somehow dethroned the human species from an exalted place in the heavens. Actually, however, the work was more theoretical than empirical, it did not explain astronomical observations any better than the Ptolemaic system, and it retained a great deal of continuity with the past, especially in the way the heavens were drawn. Copernicus was convinced that the motion of the planets took the perfect form of a circle; Kepler later showed it to be (imperfectly?) slightly elliptical.

Moreover, switching the relative position of the sun with those of the earth and moon in the solar system did not dethrone anybody.[2] After all, the center of things was not a terribly good place to be. If the earth were the center of the solar system, then the center of the solar system would be the middle of the earth. But the middle of the earth was where lava came from. The middle of the earth was where Satan lived in Dante's poem of the early 1300s, *The Divine Comedy*. If the middle of the solar system was the inferno, better known as hell—then, if anything, being away from it would be a relief.

The more radical of the two works was Vesalius's *De Fabrica*. In it, the Flemish physician drew on years of breaking the taboo of touching corpses and described what he had seen himself. Not only was the (uncredited) artwork "realistic"—in opposition to the stylized forms of extant medical texts—but it showed that the earlier knowledge of the body, passed down since Roman times in the works of Galen, was in certain respects simply wrong.

In these senses, then, Vesalius's work was more recognizably scientific than that of Copernicus. While each challenged hoary teachings passed down from the ancient world, Vesalius was able to make a case much more persuasively than Copernicus. "Do it yourself," was the central message, "and you'll see what I saw." This claim, both empirical and arrogant, was central to what would come to be known as science (although first as the "new philosophy").[3]

But Vesalius's reasonable-sounding claim was also diametrically opposed to the most basic implication of Copernicus, which was "Don't believe what you can see, for it may be an optical illusion." The sun merely *appears* to rise over one horizon, traverse a path across the sky, and set over the opposite horizon; it may really be the relative motion of the earth that makes it seem so.

What these works share is nothing particularly revolutionary or, for that matter, obvious—simply the idea that there was some new knowledge, that the wisdom of the ancestors was at best incomplete, at worst erroneous. It was in fact the kind of thing that writers have always written. If the received wisdom is complete and accurate, then what remains to be said? Presumably the reason you are writing at all is that you feel there is something that needs to be added or amended.

What made these works relevant is that they were published in a context in which authority of various kinds was already being called into question, for example by Martin Luther, and the rapid dissemination of new ideas was possible, thanks to the development of movable type.

THE NEW PHILOSOPHY

If there was any dethroning to be done, it was done by Galileo. Although a professor, he took the radical step of setting his ideas in his *Dialogue Concerning the Two Chief World Systems* (1632) in the vernacular Italian rather than in the scholarly and restricted Latin. This would guarantee the book a large audience, unless the church banned it, which they soon did.

Galileo synthesized, to a large extent, the innovations of Copernicus and Vesalius. The better of the "two chief world systems" was of course the Copernican (not the Ptolemaic or Aristotelian), but there were now autoptic ("see it for yourself") Vesalian arguments to support it. Sunspots, the moons of Jupiter, the phases of Venus—these were all pieces of evidence that Aristotle did not have at his disposal, for they became visible only with the new technology—telescopes—which in turn raised an interesting question: What Would Aristotle Do if he knew what we know now? Galileo's answer was that Aristotle would not be so irrational as to defend an indefensible proposition like geocentrism.

More important, he would abandon his old idea that the earth is stationary and that the crystalline spheres move around it.

Integral to Galileo's arguments was the nature of motion. For over a thousand years, the Aristotelians had taught that motion was a quality or an attribute: a moving marble was different from a stationary marble in the same way that a red marble is different from a blue marble. Things either possessed motion or they did not, and the earth did not.

The astronomers and physicists, however, were now thinking of motion quite differently—as a state, like being in love, rather than as a quality, like being red. To the Aristotelians, downward motion—that is to say, falling—involved being imbued with the property of gravity. Something heavy possessed more "gravity" and thus would fall faster than something light.

And yet, it did not. As the myth has it, Galileo gave a public demonstration from the top of the Leaning Tower of Pisa, showing that something light falls just as fast as something heavy. This, in turn, suggested that Aristotelians had motion all wrong. Maybe motion was indeed a state, which all things might be in at one time or another—including the earth—and as such its general attributes could be discerned. "The book of nature," wrote Galileo, "is written in the language of mathematics."

All of which was quite threatening to traditional sensibilities. Why, after all, should anyone care what language the book of nature is written in? Medieval Christianity held that this world was corrupt and evil; the *next* world was the one a good Christian should be thinking about.

This reveals the truly novel idea at the heart of the "scientific revolution"—that studying the world at all was a good thing. It was something that could not be taken for granted, a uniquely cultural assumption. Consider the writings of two of Galileo's English contemporaries: Francis Bacon and Christopher Marlowe. Bacon promoted a radical idea, that new ways of gaining knowledge would lead to new inventions and ultimately to a better life for all—in short, to progress. He begins his 1620 work, *The New Organon* (some of Aristotle's writings were collectively called *The Organon*), with a call to arms:

> Those who have taken upon them to lay down the law of nature as a thing already searched out and understood, whether they have spoken in simple assurance or professional affectation, have therein done philosophy and the sciences great injury. For as they have been successful in inducing belief, so they have been effective in quenching and stopping inquiry.

England, having recently parted ways with the Catholic Church, was one place where you could make such strong claims about the limitations of traditional sources of knowledge without too much fear (although it must be noted that Bacon did not come to a particularly happier end than did Galileo, but for principally secular reasons). The important thing is that a thirst for new knowledge of nature would lead to discoveries, whose "benefits . . . may extend to the whole race of man."[4] And who would want to argue against that?

The traditionalists had arguments of their own, however—namely, a well-founded suspicion of people for whom knowledge and power are goals. Marlowe's play *The Tragical History of Doctor Faustus,* dating from 1604, tells the story of the scholar whose lust for knowledge proves too great, so he learns magic and conjures Mephistopheles to gain the power he seeks. His first request is to know once and for all about the solar system:

> Tell me, are there many heavens above the Moon?
> Are all celestial bodies but one globe,
> As is the substance of this centric earth? (Scene VI)

And his last request involves some experimental data: he wants to smooch with Helen of Troy and find out for himself if she really was as hot as all that:

> Was this the face that launched a thousand ships?
> And burnt the topless towers of Ilium?
> Sweet Helen, make me immortal with a kiss.
> Her lips suck forth my soul; see where it flies! (Scene XIII)

And, yes, he ends up in Hell for all eternity (unlike in Goethe's later, happier version).

If Bacon and Marlowe found themselves on different sides of science—one evangelizing for it, and the other suspicious of it—their contemporary, John Donne, at least gave this movement a label: the New Philosophy, which, he added, "calls all in doubt."[5]

At stake were two fundamental issues: are there limits to what we should know, and where does the authority for what constitutes knowledge lie? The first issue is more vexing than it may initially appear, since it presumes that we can distinguish what we don't know and *should* know from what we don't know and *shouldn't* know. Wanting to know everything is a desire for omniscience, which treads heavily on God's infinitely large toes. Not only do we not want randy old geezers like Doctor Faustus possessing the ultimate knowledge of the universe today, but we also have a reminder of what happened the last time people sought such a thing. Genesis 11 tells us of the people of Shinar, who wanted to encroach on God's domain by building a tower to reach up into heaven:

> And the Lord said, Behold, the people is one, and they have all one language; and this they begin to do: and now nothing will be restrained from them, which they have imagined to do. Go to, let us go down, and there confound their language, that they may not understand one another's speech. So the Lord scattered them abroad from thence upon the face of all the earth: and they left off to build the city.

The unfinished city was called Babel. Its inhabitants were punished for their arrogance, and that is your lesson: tend your flocks and fields and aspire to goodness, but not to secret knowledge. It's secret for a reason.

And the reason is secret, too. Go back to your fields and flocks.

The second issue is more basic and more threatening. We know where to find knowledge: in the Bible. The clergy are trained to tell us what it says. Anything that isn't in there isn't worth knowing.

One of the most basic issues spearheading the Protestant Reformation was the idea that the Bible should be accessible to all, not just to the clergy. Shouldn't knowledge be democratically distributed and freely available? King James's commission indeed produced a Bible accessible to all readers of English in 1611. The ability to "look it up yourself" was intellectually empowering, just as Vesalius's call to "look at it yourself" was.

Indeed, the question was essentially rendered moot by the printing press. The ideas were now out there, in shops and nailed to doors. You could no longer prevent people from being exposed to new ideas, but you could still warn them very sternly about which ideas were bad for them. And that is exactly what the church did with Galileo's work in 1633. (As we shall see in chapter 6, when the shoe was on the other foot, so to speak, the scientific community reacted the same way in 1950 against a silly book called *Worlds in Collision*.)

The problem was that anything that was worth warning people about was also likely to pique their interest, then as now. Moreover, those who did not adhere to the Catholic faith were likely to be attracted to ideas that Rome repudiated, if only out of some degree of perversity. Perhaps that is partly why early seventeenth-century England was so taken by Galileo's banned work. A young John Milton had visited him, blind and imprisoned at the hands of the Catholic Church, in 1638, and years later wrote about Samson, blind and imprisoned in the Philistine temple of Dagon.

WHAT IS REASONABLE?

If we have knowledge as a goal, and it is fairly clear that we know more now than our ancestors did a few decades ago, then it must follow that their knowledge of the world was less perfect than ours. And yet they possessed the same sacred revelation we do. It must follow, then, that our increased knowledge about the world is coming from new and different sources—those observations and experiments that are becoming so fashionable.

Observations and experiments are reliable sources of knowledge, however, only insofar as they can be predictable and reproducible. If I try to measure the distance covered by a swinging pendulum, or the rate at which things fall (as indeed Galileo did), then the only way I can be confident that I have some knowledge is if I am reasonably certain that the experiment or measurement will come out the same way if I do it next month, or if someone else does it next year. In other words, with the growth of knowledge and the new faith placed in autoptic evidence and public demonstration comes an almost automatic assumption about how the universe works: predictably, uniformly, regularly.

The sun rises in the east and sets in the west. Always has, always will. I have confidence that the only way the sun could rise in the *north* tomorrow is essentially if the world came to an end. But that is a very specific and very demanding view of things. One could alternatively see nature as governed by capricious forces that usually operate one way but can change at the drop of a hat. Athena springs full-blown from the head of Zeus, although males cannot procreate, procreation always involves genitalia rather than crania, and all higher organisms develop and mature after fertilization. Joshua bids the sun to stand still in the sky (Joshua 10:13–14), and so we get twenty-four hours of daylight instead of the usual twelve on that day. Jesus walks on the water (Matthew 14:24), even though anything denser than water, including a person, always sinks.

These suspensions of the ordinary workings of nature are known commonly as miracles, and they began to pose a problem for thinkers who were increasingly relying on regularities of nature as sources of reliable knowledge. The more you believed in miracles, the less confidence you had in understanding how nature works, for there was always the nagging doubt that it might be pulling your leg.

Consider this exchange, once again from the Scopes trial. The evolution lawyer wants to show how irrational the creationism lawyer is, and so he asks him a series of questions about Jonah being inside the belly of a whale for three days (Jonah 2:1; Matthew 12:40). The creationism lawyer replies, "I believe in a God who can make a whale and can make a man, and can make both do what He pleases."

That is precisely the view of the universe that was becoming increasingly difficult to sustain in the mid-1600s. If the world could stop spinning for a day on cosmic whim, what about the rotational inertia that we might then expect to wreak havoc upon the earth? What about the effects of acceleration when it started up again? And if these forces could be suspended for a fairly minor occurrence like the Hebrews conquering a town, then what was the sense of studying them?

The idea that began to dominate philosophy in the seventeenth century was rationalism: that the universe operated according to rule and reason rather than by magic and miracle.[6] How else could you seek to generalize about nature unless you supposed that nature is regular and thereby subject to generalizations in the first place? Once you generalized about forces and motion, how could you not be struck by the physical consequences of temporarily suspending those generalizations?

Consequently, the dominant metaphor that came to characterize the universe in the seventeenth century was that of a great machine.[7] Machines run in predictable, regular, invariant ways. Why? Because they're built to. A really good machine may run with little tending for a very long time. A perfect machine would run forever, all by itself; all it would need was a perfect designer. The most perfect designer would be able to build a universe that could run according to his wishes, all by itself, without requiring constant interventions or tune-ups.

And we know just the Being for the job.

There was one small catch, though. If you deny God's intervention in the daily workings of the world, you naturally run the risk of being called an atheist or heretic. How could you think rigorously about the universe and its workings yet simultaneously retain the piety that would allow you to survive and function as a normal seventeenth-century European scholar? The answer was worked out by a brilliant French scholar, René Descartes, in his 1637 *Discourse on the Method of Rightly Conducting the Reason and Seeking for Truth in the Sciences.*

Like Galileo, Descartes wrote his short book in a language people could read (in this case, French) rather than in stilted academic Latin. Descartes acknowledged the need for a purely reason-bound system for understanding the universe—one based on laws, logic, and data—and set out to provide one. It begins with the only thing, he said, that he can be sure of: if he didn't exist, he wouldn't be having these thoughts. *I think, therefore I am.* From this he deduced everything else, ultimately including the existence of God. But God exists outside of the arena of bodies and movements; He is not of the world but beyond it. And what is the point of contact between the rational, corporeal, material sphere and the eternal, perfect, and divine sphere of God? The mind, the spirit, the very consciousness that permits us to know anything at all, the divine spark that differentiates us from unconscious, insensate creatures. They are "just" matter; they are essentially zombies—alive, but lacking the contact with the spiritual universe that is afforded by our soul.

Descartes thus allowed the material universe to be studied scientifically while nevertheless maintaining an appropriate degree of reverence for the spiritual universe and for the limited points of contact between the two via human consciousness. His system succeeded in establishing science (that is to say, reliable knowledge of the way the universe runs) as an achievable end and excluding the manifestations of the spirit from its domain—or, at least, relegating those to another interesting domain. Nevertheless, it also clearly underestimates the mentality of nonhuman species. In the Cartesian system, that distinction—man from beast—is the same as the distinction between spirit and matter. Man has elements of both natures; beast partakes of only one.

And that is hardly the thing Descartes was wrongest about. But Descartes' influential work allows us to steer a middle course between the faithless materialists and the defensive, backward-looking theologians. In other words, Cartesian dualism, or the mind-body distinction, allows us to sidestep the atheistic reductionism that holds us to be "just blobs of protoplasm" (which of course we may just be) and yet challenges us to query the nature of the perceptible universe in a purely rational, mechanistic fashion (which Descartes has convinced us is the only way to do it).

By the end of the century, when the English anatomist Edward Tyson publishes the first dissection of an ape, that is precisely the way he will interpret it. The chimpanzee's body is extraordinarily similar to our own—so similar, in fact, that Tyson cannot believe it was not made to walk erect as we do. So he concludes that the knuckle-walking he saw must have been due to illness. We are thus linked corporeally to the rest of the animals. What separates us from the beasts, then, is the chimpanzee's lack of articulate abilities—its inability to speak and by extension to think and feel, the glaring absence of that divine spark, or spirit.[8]

NATURE AS A MANIFESTATION OF GOD

A more radical approach was proposed by the Dutch philosopher Baruch (Benedictus) Spinoza. Spinoza began to take rationalism to its logical conclusion. Legend has it that Spinoza became an apostate from the Amsterdam synagogue after questioning how Moses could possibly have written the first five books of the Bible, generally attributed to him, when the end of the fifth book (Deuteronomy) relates his own death and burial.

In a universe of magic and miracle, Moses's feat would pose no problem. The merits of such a universe lay in its hopefulness. The problem was, is such a universe *real*? Applying the laws of reason to all things, not just to the material realm, Spinoza derived some the most retrospectively modern scientific ideas of his era. The laws of nature work the same way everywhere and always have, said Spinoza, and nature is all there is. Consequently, the division between the physical and the spiritual domains is false; it is all physical. God willed the universe into existence, but since the universe is all there is, that is what God must be—the universe itself. God is the spirit mixed in with the matter; He is everywhere and in everything.

Since God is the universe and its laws, it is silly to think that He would change His laws on a whim. Spinoza's universe thus excludes the miraculous, for what we record as miracles, says Spinoza, are simply things we don't understand at that point in time.

And what accounts for the Bible telling us about miracles? Simple: the Bible is mistaken—or at least the traditional, literal interpretation of it is mistaken. It is a historical document and needs to be understood as the product of human minds many centuries ago. To take it literally, at face value, is to miss its point. People who do so "are carrying their piety too far, and are turning religion into superstition; indeed, instead of God's Word they are beginning to worship likenesses and images, that is, paper and ink."[9]

Seeing humans as having a single nature (rather than a dual nature of body and spirit) and seeing God everywhere (rather than just in human spirit) carried an implication at once exciting and arrogant almost beyond belief—namely, that by studying the way the world works we are effectively discovering God. God is not in the Commandments, nor in the complex additional rules, nor does He intervene in the course of history. He *is* history, and He can be found in the world. He is not in the Bible; He is in the fabric of the universe. And so what will tell you more about God—studying the Bible or studying the fabric of the universe?

The latter, obviously. And this would motivate early scientists of the seventeenth century very strongly. To study nature was to study God; to do science was actually to do religion.

Consider the following assertion: "We have caught the first glimpse of our own instruction book, previously known only to God." That was made in reference to the completion of the Human Genome Project's principal goal: recording the entire DNA sequence of a human cell, by the molecular geneticist Francis Collins in 2000.[10]

What does such a sound bite convey? The human body can be understood most fundamentally as a machine, requiring a manual; it was initially the product of a Divine nature, and we have now encroached on the previously hidden domain of the Almighty; and it's worth bragging about. All three ideas would have been anathema in the fifteenth century, they were becoming mainstream in the seventeenth, and by the twenty-first they are taken for granted.

And yet, it is not clear that any of them is actually true. The first is a metaphor and thus is not supposed to be taken literally at all. It is heuristically useful to regard the body as being *like* a machine—one can learn some basic things about it that way—but it is not, strictly speaking, a machine. The division between living and nonliving things is recognized widely, and particularly in science; so, associating a body with a machine violates an obvious principle of that order and is not necessarily intuitive, much less "right."

The second presupposes some knowledge of the spiritual world, not the least of which is that there is a God, that He had something to do with DNA, and that His knowledge is now ours.

And the third implies that there is something good about colonizing God's turf, in spite of the scriptural evidence that He doesn't like it, and in spite of the value that European society has traditionally placed upon humility and upon its opposite, pride.

In short, the bedrock upon which modern science was built consists of some highly culture-bound assumptions about the nature of the material universe, its relation to the nonmaterial universe, and the value of studying it. Establishing those assumptions as normative was the greatest accomplishment of those first scientists, in the seventeenth century.

REDUCTIONISM: MACHINES AND THEIR PARTS

The powerful image of the universe as a machine that can be understood by disassembly and examination of a part at a time, revealing the mind and will of its designer, loses much of its appeal when taken too far. Modern science would say that the universe is not literally a machine but is *like* a machine in certain ways. A recent version of creationism called intelligent-design theory (see chapter 5) argues that scientists of the seventeenth and eighteenth centuries were actually right—the universe *is* literally a machine. In a weak form, this philosophy is entirely compatible with Darwinism (God built a universe in which one group of apes would survive by becoming bipedal in the late Miocene epoch). In a stronger form, however, its adherents use the assumption of a designer to infer that evolution has *not* occurred, instead envisioning a creator who continually tunes up his creation by introducing and destroying new species.

Clearly the belief in a designer of nature crosscuts the issue of whether you believe life evolved. At issue is the *kind* of machine the designer designed—one that makes biological history under its own power, or one that requires continual input.[11] This is a silly question—since the universe is *not* a machine in the first place, it is merely *like* a machine in certain ways and unlike one in others. There is no little irony, however, in the advocates of intelligent design attempting to bring cutting-edge scientific ideas of the seventeenth century into the science curriculum of the twenty-first century.

Nevertheless, the machine analogy proved fertile to early science, permitting it to reconcile the secular study of nature with the pious sensibilities of the age. God entered the picture in two places: first, as creator of the natural order, and second, as its animator. A watch, after all, was made by a watchmaker, but it still needed to be wound; and the winder is not necessarily the same person as the maker. Likewise, even if matter and motion were all there is, nevertheless matter still required an original maker, and motion an original mover.

Another way of taking the machine analogy too far lies in the failure to recognize that the artifices of human invention may be modular, but the products of nature commonly aren't. If you take a car apart and learn how the fuel injector, ignition, suspension, and other parts work, you can safely say that you understand how a car works; there doesn't seem to be anything else to a car than its parts, and you can essentially predict a car from the properties of its components. But something as simple as water defies such an analysis. Although we know water to be nothing but hydrogen and oxygen, there does not seem to be any way of predicting water from the properties of the two gases that compose it. It's not that there is something missing, but simply that water has emergent properties that arise from the combination of hydrogen and oxygen; the whole is, if you will, greater than the sum of the parts.

The idea of emergence—that increased complexity entails the production of new features that were not previously evident—runs against the idea that the universe is to be understood as a machine, piece by piece.[12] Since the universe is not really a machine in the first place, this does not pose much of a problem. Nevertheless, in the ways that the universe is indeed like a machine, reductionism is often sought and has occasionally been found.[13]

The great conceptual reduction of the seventeenth century was to reduce mechanics—the workings of nature—to regularities of two things: matter and motion. There did not appear to be a fundamental difference between the way things worked in the sky or on earth—itself a radical proposition at the time. But with the tools to describe matter and motion anywhere, provided in 1687 by Newton, you did not seem to need anything else. You could predict quite accurately where things would be and how they would act.

Such an approach was not limited to physics, either. An English anatomist named William Harvey was interested in the function of the heart and the nature of blood and brought the mechanical philosophy and experimental empiricism to bear on the subject. His 1628 *On the Motion of the Heart and Blood in Animals* (known from the abbreviated

Latin original as *De Motu Cordis*) quickly became a classic exposition of how the new ideas in mechanical physics could be applied to biology to produce new knowledge.

Harvey argued that the body was like a solar system, in a crucial way. Just as the heavenly bodies move in circles, so too does the blood. And for all the mystery surrounding the function of the heart, the nature of the blood, and tiny projections in the arteries and veins, the whole system could most productively be seen as a central pump with a system of auxiliary tunnels and valves. In 1748, a French scholar, Julien Offray de la Mettrie, would take this view to its logical conclusion, that "Man [is just] a Machine." The critical point here, once again, is that the body is not a machine; it can, however, be understood as being like a machine in certain ways.

There are other ways of thinking about things, naturally. The mechanical philosophy was especially popular among scholars in Italy, France, and England. In Germany, however, another way of thinking about nature was popular: that of smaller entities being encapsulated by larger ones. This was not about a machine comprising modular parts but about simple building blocks being compounded to form larger units. This kind of thinking was the result of the influence of a charismatic medical philosopher called[14] Paracelsus (1493–1541). Although this perspective does not shine much light on certain aspects of nature such as the organ systems in the body, it does give a useful framework for thinking about cells composing the tissues and would provide a fertile intellectual ground for the development of cell biology in nineteenth-century Germany. The organs and tissues fit together and can be understood as if the body were a machine built of functioning parts; but the cells are rather less easy to understand that way. They seem to be elementary living units, collected into larger units with emergent properties.

As the study of nature as machinery helped to demystify many of its processes, the domain of the miraculous reciprocally began to shrink. In biology, however, the nature of life did not readily yield to mechanistic analysis. Those things that had once been alive or had never been alive indeed seemed to be fundamentally different from those things that *were* alive. Moreover, there was a strange connection between the dead and the living: dead meat could produce living maggots. While there is nothing particularly Biblical about it, the doctrine of "spontaneous generation" taught that dead matter could be straightforwardly transformed into living matter; could not spontaneous generation be seen as God's divine workings on earth, recapitulating the miracle of the first creation on a small scale every day? Indeed it could, said the pious English scientist John Turberville Needham in the mid-1700s.

Nonsense, replied the French *philosophe* Voltaire, who had use for neither a theory of spontaneous generation nor a theory of an actively intervening Deity. Spontaneous generation would in fact not finally be put to rest until the middle of the nineteenth century, and the scientific problems caused by doing so (all right, so where does life come from and how is it passed on, then?) would be reconciled only with the development of the cell theory: that cells are the units of life, and all cells arise from preexisting cells. Now, to the extent that the spontaneous production of life could be considered miraculous, those particular miracles were clearly not happening regularly anymore, and—who knows?—life as we know it today might even simply be the end result of as few as one single transformation from nonlife to life, a very long time ago. That indeed would be the very last thought expressed in *The Origin of Species* in 1859—that life had been "originally breathed into a few forms or into one."

THE IMPACT OF AIR PRESSURE

Perhaps the most successful early exponent of the experimental and mechanical philosophies was the seventeenth-century Englishman Robert Boyle, who was interested in the physical aspects of a question with strongly metaphysical implications.

The ancients had taught that nature abhors a vacuum. A vacuum is the absence of anything, and the universe is filled with something. Could people create a vacuum? If so, what would that imply? If a vacuum is nothing, isn't nothingness hell? Would a vacuum be a portal to Hades?[15]

By the mid-1600s, however, two sets of experiments had converged on the ideas that (1) vacuums are indeed possible; (2) they are not a portal to anything, just the absence of air; and (3) air is not what you think it is. One set of experiments involved the invention of an air pump, by Otto von Guericke, which he used to suck the air out of a two-part brass sphere. When the air was out of the sphere, a team of horses could not separate the two halves of the sphere from one another. This was hard to understand, unless you supposed that air had some kind of force that pushed against things and the inside of the sphere had nothing pushing back.

The second line of evidence came from an invention by the Italian physicist Evangelista Torricelli. Filling a closed-ended tube with the heavy liquid mercury and then inverting it in a pool of mercury, you always saw the mercury in the tube settle, leaving some space remaining at the top of the tube. Yet you had not let any air into the tube, so what was in the space? Torricelli had in fact invented the barometer, which we now know to be sensitive to air pressure, but how could air have pressure?

These were the questions that inspired Robert Boyle in London.[16] Independently wealthy, he outfitted a laboratory and employed technicians to build the apparatus that would permit the emptiness that had not previously existed on earth to be studied on a small scale, under rigorous and controlled circumstances. Assuming that there was a transparent relationship between the small-scale man-made phenomena that he could examine and the large-scale natural phenomena that they ostensibly represented, he could come to understand the universe a little bit at a time—and in public, for any and all to see. This system of knowledge would become a foundation of modern science: representing long-term, large-scale natural processes in a man-made microcosm. The burden of proof would now subtly shift from the claimant to the doubter, to try and identify a crucial mismatch between the experimental conditions and natural processes, which would invalidate the extrapolation from laboratory to cosmos.

Boyle's experimental philosophy of physics and chemistry would provide a basis for Darwin's argument in *The Origin of Species* two centuries later. Darwin would invoke the diversity of pigeon races created by breeders over the recent course of human history as evidence that some similar kind of selection by nature, over the long course, could also produce a diversity of species. As Boyle himself wrote, in the midst of defending the new experimental philosophy, the argument is not over what God can do but what nature can do.[17]

Nature may not like a vacuum very much but did not seem to abhor it, despite received wisdom to the contrary. Boyle, after all, was able to produce a vacuum on demand. Blaise Pascal had shown that, whatever the barometer read, there was less of it on the top of a mountain than at sea level; and Boyle himself showed that a barometer inside a vacuum chamber hardly registered anything at all. It was not so much the weight of air—the difference in weight was negligible—but another property of air, a force like a spring's. It was as if air were composed of invisible particles bouncing around, which bounced harder (and thus offered more resistance) when forcibly compressed and more weakly when given more space.

This was, to say the least, nearly as counterintuitive as maintaining that the earth goes around the sun when you can plainly see the sun rise, cross the sky, and set over the opposite horizon. Now we were supposed to believe that air was composed of tiny invisible things that could push.

And yet the tools that interpreted air as invisible pushing things, and the earth as being in motion around the sun, appeared to work very well in predicting the behavior of things under experimental circumstances. The experiments could in theory be viewed by all but of course in practice were not. All interested parties could not arrange to be there to see them at any one time, nor could most interested parties afford to build their own apparatuses. What emerged was the development of a form of "virtual observation," in which someone, say, Boyle, would describe in an abbreviated form what he (or his assistants) actually did, and what you would have seen if you had been there, or would see if you could do it yourself, as well as what it means—in other words, the scientific paper.

And yet, the scientific paper required an infrastructure before it could be an effective means of disseminating knowledge. Not only did there need to be a critical mass of people interested in reading such a document, but there needed to be the financial structure that would permit the knowledge to be produced and also subsidize the publication of that knowledge. This necessitated some kind of patronage—the richer, the better—as indeed Galileo had courted the Medicis by naming stars for them at the beginning of the century.

Wealthy patrons began to promote the production, dissemination, and discussion of the new knowledge. Generally acknowledged to have been the first scientific organization was the Accademia dei Lincei (Academy of the Lynxes), incorporated in 1605 in Rome, whose members self-consciously compared their ability to see things clearly with that of the sharp-eyed cat. By mid-century there was an "invisible college" operating in England, both complementing and to some extent competing with the visible colleges for the domain of what counts as authoritative knowledge. In 1662, it became the Royal Society of London for the Promotion of Natural Knowledge, patronized (although not actually subsidized) by the crown.

Two other frameworks also helped make scientific knowledge authoritative. A standardized system of weights and measures ensured not only that business was transacted fairly but that experiments could be (at least in principle) accurately translated from place to place. The relationship between the two—the development of science and public metrology[18]—was so intimate that Isaac Newton would spend his postscience years in the early 1700s overseeing the standardization of English money as Master of the Mint.

The other framework also involved the participation of the public. Although its principal practitioners were the idle rich, the clergy, and the universities, science was actually subversively democratic. It relied not upon God-given talents or gifts, but on the simple application of a proper method or system. Although the particular system might be in dispute—whether Bacon's, Descartes', or someone else's—it was something that essentially anyone could do with the appropriate training of the mind. Genius was nice, of course, but not necessary. Moreover, a system of formally assigning credit (and profit) from new truths, or their applications, made it a possibly rewarding endeavor.

The implications for practice were revolutionary. In medieval times, when inspiration or genius was thought to be a gift from beyond, it seemed improper to take too much credit upon oneself or to profit from it materially. Indeed, it seemed uncomfortably close to simony—the sin of monetizing God's gifts, after Simon Magus, who tried to purchase the divine experience in Acts 8:19.

But if science was not a gift, but rather was the application of a set of rules and procedures, then presumably anyone could do it, and presumably anyone could also benefit materially from its practice. Scientific discoveries were thus to some extent inevitable, and consequently what mattered was not so much who did it, but who did it *first*. The crystallization of a scientific community went hand in hand with the development of a premium on priority. Science's greatest embodiment—Isaac Newton—would also be driven to obsession by his claims of priority—against Robert Hooke, Gottfried Wilhelm Leibniz, or anyone else who trod too close to things he had worked on.

THE APOTHEOSIS OF ISAAC NEWTON

It has often been observed that Isaac Newton was born the year that Galileo died (1642). Newton died in 1727, a year in which apparently nothing else particularly interesting took place. In between, he became the first superstar of science, the man against whom all future practitioners would be judged, and the cultural icon of his age. He was Einstein and Elvis in one. On Newton's death, Alexander Pope wrote a couplet in his honor: "Nature and Nature's laws lay hid in night: God said, 'Let Newton be!' and all was light."

To understand Newton's achievement, we have to appreciate a fundamental intellectual dichotomy that existed in the study of nature in the seventeenth century. On the one hand there was *natural philosophy,* the study of causes and of generalizations about why things are as they are. That term is now archaic. On the other hand there was *natural history,* the study of the particulars, the patterns and expressions of nature, and a phrase that is still with us. The difference between them was essentially the difference between philosophy and history: the analysis of why things happen and what they mean versus the documentation of what actually happened and what is out there.

Newton, from middle-class origins, was an average student who showed exceptional talent in mathematics and impressed his math professor at Cambridge to such an extent that he was named his professor's successor at age twenty-six. He worked hard, had no romantic entanglements, and developed a reputation as being a brilliant (if weird) mathematician.[19] He had solved several vexing problems about matter and motion, which he had told people about but not published. Finally, Edmund Halley (who was busy studying the periodicity of comets) prevailed upon him to publish his work as a book and even underwrote its publication. The result in 1687 was *Mathematical Principles of Natural Philosophy* or, simply, Newton's *Principia*.

It has been estimated that fewer than one hundred of his contemporaries actually read the book. Nevertheless, the scope of its impact was astounding, because Newton marshaled the rigor of mathematical proof in support of generalizations about motion and gravity. Not only had Newton produced these laws, but he had derived them using mathematical tools he invented himself and called "fluxions" (we now call it calculus), though he presented his proofs in the older language of geometry.

What Newton seemed to have done was to have united the celestial and the terrestrial spheres with a single theory that explained the motion of the planets as well as the motion of apples. Moreover, he explained gravity as an attraction between any two objects—sun and earth, or earth and apple—in a way that unified not only heaven and earth but anything and everything else. The most basic properties of the cosmos—being and moving—could now be decoded, as it were, and the rules that govern them could be learned. *Why* those rules exist as they do, he couldn't say—and indeed wouldn't say: I proffer no hypotheses ("Hypotheses non fingo"), he wrote.

Newton divided his energies among mathematics, alchemy, astrology, and Scripture, and in fact rather few of his writings were actually about science. As a result, the economist John Maynard Keynes could judge that "Newton was not the first of the age of reason: he was the last of the magicians." But Newton's impact lay in circumscribing the supernatural realm and shielding the natural realm from its effects. This is not to say that there were no mystical elements in his system—Newton's gravity is an invisible force of attraction between two bodies, after all—but the force was law-bound, not capricious.

Finally, the *Principia* synthesized the major tensions in the nascent scientific community. Like Boyle, Newton relied on some experimental data; and, like Descartes, he produced a mathematized system of the universe. Like the natural philosophers, he explained why things work as they do, and like the natural historians he explained the facts of nature.

By compartmentalizing or segregating God and His realm apart from His creation, Newton was making a contribution not only to science but to religion as well. He ventured no hypotheses about how God went about His business of creating, and he did believe that the universe required some intervention now and again. But others went further than Newton and called attention to the theological problem introduced by the new scientific system of nature.[20] A self-sufficient universe is better than a universe that requires a constant input of miracles to set it straight. If God created the universe and is perfect, then shouldn't the universe now be free of His attention? God's magnificence and creativity could hardly be attested by an imperfect universe, and it therefore followed that the universe must be perfect—a conclusion famously satirized in Voltaire's *Candide.*

More to the point, though, the universe must be devoid of God's intervention, since a universe created by a perfect Being would not necessitate such intervention. The Newtonian universe—if not Newton's personal universe—was thus free of miracles and of God's presence. God had been there when we really needed Him—at the beginning—but everything since then had occurred according to the rules He set up, which we were now in the process of discovering. (This also entailed problems for free will, needless to say—were all of your decisions laid down at the beginning of time?—another consequence of overapplying the analogy to machinery.)

THE VIEW FROM NOWHERE

As the goal of the new route to knowledge—reason and experimentation—came to be valued more highly, and Newton refrained from mixing the theological with the scientific, the best-known scientific norm (or myth) began to take shape. This was the idea that science should be free of not just theology but any ideology whatsoever. Politics, values, ethics—all were to be bracketed off from science. The ideal scientist would have what the modern philosopher Thomas Nagel called "the view from nowhere"—that is, a completely open mind, free of any prejudice or perspective.[21]

And yet, from its very inception modern science was used to underpin political ideologies. Thomas Hobbes crafted an ostensibly scientific argument in *Leviathan* (1658) to prove the necessity of a monarchical government. A few decades later, John Locke would even invoke Newton for his philosophy of knowledge in *An Essay Concerning Human Understanding* (1690), and his influential vision of people as having "natural rights" would be possible only in the new conception of nature as something real, regular, and divinely ordained. Over a century later, Karl Marx would frame his ideas not as socialism but as *scientific* socialism. Indeed, the very label "political science" would seem to be an oxymoron if science and politics are taken to be discrete and unconnected realms.

The idea that science should be dispassionate and apolitical is one of the most interesting assumptions about it, since in fact science has never been either of those. After all, passion—that is to say, unchecked obsession beyond the bounds of reasonable behavior—is one of the hallmarks of the successful scientist in any age. Science is not, and has never been, a nine-to-five job. Moreover, as Steven Shapin points out, Francis Bacon, the apostle of modern science, in seeking acceptance and patronage for science, was explicit that scientific knowledge is empowering to the nation, and the nation should control that knowledge.[22] "Knowledge," Bacon wrote epigrammatically, "is power."

At the same time, however, Bacon was promoting science as the best path to reliable knowledge, particularly because it was free of the intellectual prejudices that other systems possessed. In particular, Bacon saw himself as a second Abraham, a true iconoclast, smashing the idols worshipped by false systems of knowledge. Bacon identified four such idols: Idols of the Tribe (falsehoods rooted in a lack of fit between nature and the ability of the human mind to perceive or render nature accurately); Idols of the Cave (falsehoods rooted in unquestioned cultural assumptions); Idols of the Marketplace (falsehoods rooted in the use of language); and Idols of the Theatre (falsehoods rooted in dogmatism). Bacon argued that a true system of science must free itself from such idolatry.

Of course, it is considerably easier to identify someone else's intellectual prejudices than to identify your own. What Bacon seems to have started is not so much the quest for unprejudiced knowledge but the self-delusion that a scientist can actually be free of the "idols." If there are falsehoods that result from the structure of the human mind, then the only way to avoid them is not to be human, which is patently impossible. If there are falsehoods that result from culture, language, or unquestioned adherence to assumptions, how can a scientist possibly be free of them, short of carrying out the scientific activity in another language, in another culture, and with a basically anarchic frame of mind? Bacon argues for precisely the opposite—a rigid system for truth determination—and most important, *his* system. And all, un-self-consciously, for the glory of his own society and government.

What's striking is thus not so much the ambition but the hypocrisy, or lack of reflection, that accompanies the ambition.

Not only has science been linked to political power from the beginning, but far from being in a constant battle with religion it has usually been wed to religion. This association runs a gamut: On the one hand, there is a trivial and subtle association in people tending to be religious and scientists tending to be people. On the other hand, scientists have commonly interpreted their work in an overtly religious frame. Natural theology, the biology learned by Darwin in college, was explicitly understood as the study of God's wisdom and bounty.[23] Darwin's contemporary, the French philosopher Auguste Comte, envisioned a progression of human intellect that would later crystallize into a rise from superstition or magic, through religion, and then up to science—

science as destined to replace religion, as its new and improved mode of thought. In other words, there is something primitive about retaining religious beliefs and attitudes when scientific beliefs and attitudes are the ones that are truer.[24]

A modern version of this doctrine is that of the biologist Richard Dawkins, who is led by his interpretation of nature to the conclusion that God does not exist and who is famously scornful and condescending toward anyone who may be led in the opposite direction.[25] Far from being a "view from nowhere," however, his is a view from somewhere, and a very self-interested somewhere, to boot. After all, you couldn't expect a scientist to be more candidly humble about the limitations of science than you could expect a Pontiac salesman to be about the limitations of Pontiacs—or, for that matter, a Jehovah's Witness to be about their beliefs. It is directly in these folks' interests to have you accept their spiel: they either set themselves up as the voice of authority, sell a car, or win a convert.

Actually, the "warfare between science and religion" is largely a product of the late nineteenth century—a very modern idea. It was in large measure declared by Thomas Huxley, who never hesitated to explore the possible social, religious, and philosophical implications of science.[26] Moreover, with the goal of professionalizing science education, he actively sought to delegitimize the local clergymen who were teaching science to most of England's schoolchildren.

The idea that science is locked in a perpetual struggle with religion is a bit extreme. It relies on a fairly narrow definition of religion and a credulous approach to science. The two may come into conflict, particularly in the narrow zone that houses origin narratives, but it seems unlikely that science will, or should, replace religion. This is a theme to which we will return in subsequent chapters. For the present, simply reflect on this scenario: You've just done something you should not have. You feel a burden of conscience, perhaps even shame, and you feel the need to confront the feelings of remorse and make restitution or atonement. Whom should you call—a minister or an X-ray crystallographer?

THREE Normative Science

The practice of science, like any other social endeavor, entails rules of conduct. Commonly these rules are not explicitly articulated but are assimilated by participation in the community. Although the question "How do we do science?" was answered in various forms by Francis Bacon, René Descartes, and others from its inception, the modern philosophy of science essentially begins with a group of philosophers clustered in Austria in the 1930s. The "Vienna Circle" maintained (as had William Whewell, who coined the term *scientist* in the nineteenth century) that the purest and most advanced sciences were physics, math, and astronomy, and they were consequently the sciences that should be emulated. Quantification, generalization, being able to discover broad regularities, as the scholars of the seventeenth and eighteenth centuries had done—*that* was science, and everything else was a pale imitation.

SCIENCE AND TRUTH

Karl Popper began his career under the sway of the Vienna School but rejected two of its central tenets. The physical sciences were indeed the avatars of true science, but it was not so much their generalizing or mathematical aspects as their experimental aspects that held the key to understanding science. Moreover, while mathematicians and astronomers were sometimes able to establish physical generalizations based on repeated observations, the exceptions in nature and the laboratory made it difficult to see establishing empirical or mathematical regularity as the key to science. Rather, said Popper, science proceeds principally by proving statements false, not true.

Elaborating a critique of inductivism—the idea that science begins, or should begin, by the open-minded and unprejudiced collection of data which can then be assembled and analyzed for its patterns—initiated by the eighteenth-century Scottish philosopher David Hume, Popper acknowledged that no data collection is unprejudiced. Simply a decision about what is worth recording is prejudice.

As an example, consider the famous story of Darwin's being impressed by the finches on each particular Galápagos island and deriving inspiration thereby for his theory of natural selection. In fact, however, Darwin went to the Galápagos in 1830 as a creationist, and with an idea about adaptations. Since God had adapted animals to their environment, and the environments of the different islands of an archipelago were identical, it followed that however diverse the finches looked, it didn't really matter which island any particular specimen came from. So he did not bother to write that information down. On studying Darwin's collection of finches from the Galápagos, the ornithologist John Gould asked Darwin about the provenience of each finch specimen, and Darwin then had to reconstruct it, largely by matching up his scientific specimens against specimens that the sailors on the voyage had brought home, and for which, not being scientists, they had recorded the irrelevant information about where the birds actually came from.[1]

Science does not begin with the objective collection of data. That would make it a mindless, disorganized activity (like sailors collecting birds). Science begins first with the perception of a problem, and then with a decision about the kinds of information that might be appropriate to solving it—both of which are ways that theory, from the outset, constrains and channels the collection of data.

What distinguishes science from metaphysics, then, is not the amassing and marshalling of evidence, but the ability to distinguish among alternative possibilities and to eliminate some of them. The elimination comes in the form of matching the hypotheses up against the real world in some manner and filtering out the hypotheses that aren't compatible with it. In this way, science progresses through a process of selective elimination of ideas—Popper called it conjecture and refutation, and the Nobel laureate immunologist Peter Medawar called it, more euphoniously, proposal and disposal.

This encodes a crucial paradox about science, however. If scientific activity proceeds by falsifying hypotheses, and if only a small minority of hypotheses are actually *not* false, then it follows that most ideas most scientists have, or have ever had, *are* false. They have been falsified, disposed, discarded—and yet are no less scientific for it. Thus, the work of Karl Popper divorces "science" from "being right"—and while formalizing its method also erodes its authority.[2]

Popper, an émigré to England, was knighted in 1965, and his sway over the philosophy of science was so enormous that science came to define itself as Popper had defined it, and nearly everything in the field subsequently has come as a reaction to his work. At one time he sent a shock wave through the field of biology by suggesting that Darwinism was not scientific, a position he later distanced himself from.

Popper's role is similar to that of Sir James Frazer, the great premodern anthropologist. Frazer, whose classic work, *The Golden Bough,* was first published in 1890 and remains in print today (although in abridged form), was a brilliant and insightful analyst of the religions of the world. What he lacked was a knowledge of—or much of an interest in—what real people thought and did. Coming up with a grand synthesis of their common themes across diverse religious beliefs, Frazer was ultimately undone by ethnography, which treated people as rational actors who did things for their own reasons—and not as mere automata.

Popper's scientist comes into the lab each day, takes a couple of sips from the glass of truth, and disappears for the evening. In the same way that Frazer was eclipsed intellectually by Bronislaw Malinowski, who defined modern anthropology as the study of what people really think and do, so too was Popper eclipsed by Thomas Kuhn, who likewise introduced the realities and complexities of human behavior into the study of science.

SCIENTIFIC REVOLUTIONS

Kuhn's 1962 book, *The Structure of Scientific Revolutions,* relegated the kind of science Popper focused upon to history's footnotes. While indeed scientists do, by and large, advance our knowledge by falsifying hypotheses, that is not how science itself really advances, argues Kuhn; that is just the nuts and bolts of it. The real intellectual activity comes in trying to make sense of the data that don't fit. As these exceptions accumulate, a new intellectual framework—or paradigm—for understanding the data emerges.[3]

But the cost is high. The emergence of a new paradigm creates a crisis, for both paradigms cannot coexist easily—or, at least, their adherents cannot coexist easily with each other. Taking a cue from anthropology, Kuhn argues that scholars adhering to different paradigms cannot even argue sensibly with one another, for not only are they thinking about things differently, but they are also using key terms in distinctly different ways and can only talk past each other. Neither paradigm necessarily explains things better, but they do expose different questions and engender the collection of different classes of data. The arguments between the exponents of different paradigms may be resolved not so much by empiricism as by the skillful exertion of social power. Indeed, the adoption of a paradigm may not be rooted in the careful evaluation of the arguments for it and their implications but in its aesthetics or in a flash of insight. That is to say, while the day-to-day activities of normal science—Kuhn's term for the activities of scientists collecting data and testing hypotheses—are very rational, the tension between competing paradigms contains significant aspects of irrationality.

But the rational aspect of science, the normal science, is boring. Most normal science is never read and never cited. The activity is tedious and is known to insiders somewhat disparagingly as "turning the crank." What is interesting about science, the intellectual ferment, is encountered at the conflict zones between paradigms.

We may consider a few examples from biology. Prior to the eighteenth century, the diversity of species was generally interpreted linearly, as falling on a line leading from the lowest forms up to the highest, that is to say, up to the human species. This has come to be known as the Great Chain of Being.[4] The Swedish naturalist Carl Linnaeus introduced a different way of thinking about natural diversity, not in terms of how similar species are to humans, but in terms of how similar species are to one another. When species were seen in this way, a very different pattern emerged: they fell readily into natural clusters of greater or lesser size and exclusivity. There appeared to be, say, vertebrates, mollusks, insects, and worms,[5] which appeared to be built according to different plans and could hardly be compared with one another, much less ranked along a single scale. And within the vertebrates, there were fish, amphibians, reptiles, birds, and mammals. One could certainly argue about the details of the categories, and certainly the qualities of these different kinds of creatures had been noted as far back as Aristotle, but the fundamental existence of those categories was obscured by the interpretation of species as constituting a single chain. When Linnaeus introduced his system of nested categories for zoological and botanical classification, it swept through eighteenth-century academe. Why? Because it was just so obviously right. The lone significant holdout was Comte Georges-Louis Leclerc de Buffon, an eminent French naturalist who considered the implications of the Linnaean system appalling:

> And if it is once admitted that there are families of plants and animals, that the donkey is of the horse family, . . . then one could equally say that man and ape have had a common origin like the horse and donkey. . . .
>
> The naturalists who establish so casually the families of plants and animals do not seem to have grasped sufficiently the full scope of these consequences, which would reduce the immediate products of creation to a number of individuals as small as one might wish. For . . . if it were true that the donkey were but a degenerated horse—then there would be no limits to the power of nature. One would then not be wrong to suppose that she could have drawn with time, all other organized beings from a single being.[6]

Linnaeus was interested in the patterns of nature, without regard to the processes that produced them. Buffon could not conceive of divorcing the two, and in turn saw the study of biological pattern without regard for process as the depths of anti-intellectualism. The two simply had to be understood together, and how could scholars fail to ask what produced the pattern they observed or inferred? Buffon recognized that there are some very profound consequences of the nested Linnaean hierarchy, like the possibility that it reflects a common ancestry of the diverse species being classified, a notion so heretical as to be essentially unthinkable, if not downright idiotic.

How could you even argue about natural patterns if you couldn't agree that pattern and process had to be related in some significant way? How could you argue about an optimal classification if you couldn't agree on whether such groups even exist? How could you argue about "mammals" if you couldn't agree on what such a cluster of animals might signify—much less whether to include a dolphin or a platypus within it?

But acknowledging the existence of the nested Linnaean hierarchy also necessarily opened up a host of other questions, which had been largely invisible before then: Were all species produced at the same time, or at different times? Could new species come into existence? If so, how? (Linnaeus acknowledged toward the end of his life that hybridization could produce new species of plants.) And—perhaps begrudgingly—why does the pattern even exist?

For a century, scholars read Buffon's *Histoire naturelle* (one of the best-known works of the French Enlightenment—all forty-four volumes) and also adopted the Linnaean system. Both naturalists were right, both were wrong, and their views were entirely incommensurable with one another. It was not until Darwin's work in 1859 that the process which generated the Linnaean system—common descent—was finally appreciated and Buffon's worst fears were actually realized.

As another example, we may consider the study of heredity from 1865 to 1900. There were, obviously, many scientists interested in the problem of how offspring come to resemble their parents. They began by assuming that the transmission of features and the development of those features were parts of the same problem. Charles Darwin proposed a model that incorporated the transmission of little buds from each body part into the reproductive organs, which then somehow merged and grew into their respective parts in the child. Breeding pea plants to study hybridization, Gregor Mendel published his results in 1865, showing that, if you (1) divorced heredity from development and just focused on the former, and (2) considered the transmission, not of body parts, but of instructions *for* the parts, you could see certain regularities in their transmission. But there was no compelling reason to believe that either of those assumptions held widely, if at all—and his work remained appropriately obscure. Over the next thirty-five years, however, the development of cell biology made both of those propositions more tenable and even introduced a third proposition: that since cells were the basis of all life and encapsulated the mysterious mechanism of heredity, heredity could work the same way in all life.

The idea that heredity in peas or fruit flies might throw light on the processes of heredity in humans is counterintuitive from the start. Peas, flies, and people are all quite different, after all. While similarities could of course be noted between the reproduction of domestic animals, say, and people, it was easy to write them off as casual, superficial resemblances, like seeing the shape of a camel in a cloud.

What, then, of the paradigm shift in the study of heredity? By the turn of the twentieth century, the intellectual conditions that had rendered Mendel's work exceptional and marginal had changed, by virtue of the rise of cell biology. Cell biology had created an intellectual space within which Mendel's work made sense. Studying (1) heredity without development, (2) bits of information (rather than traits themselves), and (3) diverse species (or model organisms, unified by their common cellular processes) led biologists such as Hugo de Vries in Holland and William Bateson in England (as well as Carl Correns and Erich von Tschermak) to rediscover the regularities Mendel had seen and to acknowledge that Mendel had been there first—and quickly transformed the study of the peculiarities of heredity into the formal generalizations of genetics.

The transformation was swift and total,[7] although none of the three central tenets had been proved but, rather, assumed. Indeed, a century later, (1) the field of "evo-devo" has emerged from the assumption that development and heredity need to be put back together, particularly in the context of evolution; (2) the conception of genes as "information" or "instructions" is now recognized to rely too heavily on linguistic metaphors and not enough on the coproduction of the organism from the genes and the cellular matrix in which they exist, as well as the extracellular conditions of growth; and (3) model organisms also have their limits: male humans have crossing over but male fruit flies do not, female humans with only one X chromosome have Turner's syndrome but female mice do not, and the primroses studied by Hugo de Vries have their own peculiar chromosome rules.

The point is that the rapid emergence of the field of genetics in the early twentieth century, including William Bateson's naming it in 1906, owed little to any single discovery or succession of discoveries, and particularly not to those of Gregor Mendel. Rather, it was the result of a lot of normal science, catalyzed by the new way of thinking about things afforded by cell theory.[8] The result was a novel and spectacularly successful set of research strategies for studying heredity, although the basic assumptions upon which it was constructed are by no means universally true, and its generalizations are commonly more honored in the breach (including linkage, crossing over, mitochondrial inheritance, pleiotropy, epistasis, imprinting, codominance, and quantitative traits).[9] Indeed, textbooks of the early twentieth century diverged according to those that made Mendel's work the central organizing principle of the study of heredity (as we do today) and those that relegated it to an interesting sidelight, meriting a chapter.

But the Mendelian way of thinking about things, and the attendant questions answered in its research program, turned out to be immensely productive, especially in the hands of Thomas Hunt Morgan at Columbia, and those of his students, working on the fruit fly, *Drosophila melanogaster.* It was their writings that crystallized Mendelian theory as the centerpiece of genetics and even codified it into two "laws"—Segregation and Independent Assortment—not previously acknowledged as such either by Mendel himself or by the first wave of Mendelians.[10]

To Kuhn, then, the interesting aspect of the history of science lies not so much in the collection of data and testing of hypotheses ("normal science") as in the proposal and ultimate adoption of new intellectual frameworks for making sense of things ("revolutionary science"). These paradigms serve a rational intellectual function but are commonly adopted for intuitive, aesthetic, or other nonrational reasons, yet the "paradigm shift" is quite possibly the stuff of the major advances in science.

Unfortunately, these paradigm shifts and scientific revolutions are generally visible only in hindsight. That doesn't make them less real, only harder to identify, especially at any point in time. After all, for every Newton, Boyle, Darwin, Freud, Einstein, or Boas there are ten thousand Joe Blows (sorry, *Doctor* Joe Blows) and even more adherents of meme-ology, numerical taxonomy, catastrophe theory, evolutionary psychology, chiropractics, postmodernism, and parapsychology, not to mention scientology and intelligent design—all interesting intellectual fashions of one sort or another, none of which is likely to be mentioned normatively in college textbooks twenty years hence. Worse still, the implicit value judgment between the dull data collection of "normal science" and the intellectual vitality of "revolutionary science" means that all practicing scientists who have heard of Kuhn's work envision themselves as participating in a paradigm shift rather than as being just another cog in the wheel of normal science.

Kuhn's great contribution was his very anthropological application of cultural relativism to the study of science. The clash of paradigms is a clash of cultures—and not in the minor sense of "you say po-tay-to, and I say po-tah-to," but in the sense of "you say potato, and I don't even acknowledge that we are discussing an edible tuber." A later generation of scholars, influenced by Michel Foucault, would unify Kuhn's incompatible scientific paradigms with the fundamentally incompatible cultural *worldviews* from which they were originally derived and call the specific, underlying assumptions about how the world works that are unique to any community of thinkers, scientific or not, *epistemes*.

OTHER REACTIONS AGAINST POPPER

If Kuhn showed that a strict application of Popper's ideas about science failed because they did not show how science really does progress, the Hungarian-born philosopher Imre Lakatos showed that they failed as well for not taking into account how scientists really think. Scientists aren't robots; they have a lot of investment in their ideas and are not about to give them up so easily. If one scientist claims to have refuted another's ideas, does the second scientist take early retirement? Of course not, he tries to find a way to protect the theory. And, after all, couldn't one argue that the crucial experiment which appears to falsify a hypothesis is what is flawed rather than the hypothesis itself? You bet one could.

Moreover, in practice, scientists rank their working ideas and sacrifice the less important ones to "tweak" the system into compatibility with the data. All of this adds up to considerably more ambiguity in "falsifying a hypothesis" than it initially seemed. Two well-known examples may suffice here.

In the 1860s, Darwin proposed a theory of heredity, called pangenesis, in which each body part secreted small templates of itself (gemmules) into the body's fluid, which ultimately passed into the reproductive organs and emerged in the sexual fluids associated with conception. His cousin Francis Galton carried out an elegant experiment. He transfused blood from one color strain of rabbits into the veins of another color strain of rabbits, then mated the recipients. Would the offspring look like the strain should, or like the ostensibly transfused gemmules? Answer: like the parental strain, not like any transferred gemmules from the other rabbit. Faced with what appeared to be a colossal refutation of this particular theory, Darwin responded rather lamely, "I have not said one word about the blood." There were, after all, other circulating fluids that might be transporting gemmules.[11]

A century later, population geneticists were vexed by the incompatibility of two models describing the nature and kinds of genetic variation that exist in natural populations. One model, the "classical," held that most people had two copies of the normal version of each gene and one copy of a few rare variants, with selection mostly acting to weed out pathological changes. An alternative, the "balance" model, held that most people had two different versions of most genes (there was consequently no "normal" allele) and that selection acted mostly for diversity, as in the well-known vigor of hybrid stocks of domestic plants and animals. The technical ability to study the amount of diversity in genetic systems became available in the 1960s, and it quickly became clear that natural populations had lots of genetic diversity. This seemed to be more compatible with the balance theory than with the classical, but did the adherents of the classical theory give up? Of course not. They modified their position, arguing now that the large amount of variation detected was largely meaningless—or neutral—to the survival of organisms, and proclaimed victory.[12]

Not only did Popper not take into account the psychological and rhetorical aspects of scientific refutations, but he overestimated the ability of science to refute hypotheses unambiguously. There is always wiggle room, and there is always strong motivation to try to find it. As a result, although the act of hypothesis testing and rejection certainly occurs, it is generally by no means as straightforward and clear as it may sound.

Further, there is a subtle but critical difference between testing hypotheses and framing consistency arguments. It is very common to encounter scientific arguments that proceed, "I found A, therefore I think B." Granted you found A, and granted that you think B, how exclusive is the connection between the two? How much less compatible with A is C, the hypothesis you don't believe? Does A necessitate B by falsifying C, or does A simply make sense in the context of B while also making sense in the context of C (although you neglected to mention it)? The former is hypothesis testing, and the latter is a consistency argument. Hypothesis testing is what differentiates science from other forms of intellectual activity; making a consistency argument is not—it's what everyone, from lawyers to shamans, does: they try to convince listeners by adducing evidence.

And that is the sense we must make of the following assertion by a well-known biologist: "The universe we observe has precisely the properties we should expect if there is, at bottom, no design, no purpose, no evil and no good, nothing but blind, pitiless indifference."[13] But perhaps, in this case, we may interrogate the scientist's ability to tell a benign universe from an indifferent one. Couldn't things conceivably be worse, and wouldn't that imply that the universe is not as horrible as it could be? It is as if the scientist were looking at a cuneiform tablet and remarking that this is just what *Hamlet* would look like if it were written in cuneiform. And it rather is, but that is a statement about your inability to make sense of cuneiform, not about *Hamlet*. After all, that clay tablet is also what *TV Guide* would look like to you if it were written in cuneiform. The issue is your ability to tell them apart, to distinguish among alternatives, and to falsify some. Whatever the overall limitations of Popper's system, this distinction is a crucial one. If you lack the discretionary tools to be able to tell an indifferent universe from a benign one, then the two are effectively the same to you. One bit of gibberish is equivalent to another bit of gibberish.

So what we have there is not a scientific proposition about the nature of the universe but—a bit more creepily—a nonscientific proposition being camouflaged by a scientist to look like a scientific proposition. That act has a venerable history and is a significant downside of reducing the scientific process to a simple formula.

The most fundamental statement we can make about science stretches directly from Newton to Popper: that science involves building a wall around the physical universe, or the natural world, and shielding it as far as possible from the universes of values, morals, and spirits. These other universes are quite possibly more important than the natural one—we trust scientists to design weapons but not to decide whether to use them.

But even that last thought contains some frightening implications. Why shouldn't scientists be involved in the decision to use (or not to use) weapons?

Short answer: because they're not trained in those decision-making capacities and don't have a particularly honorable history in that arena.[14]

Scientists themselves are moral actors, even if they separate their subject matter from the subject matter of moralists, theologians, and politicians. Two major episodes from the mid-twentieth century established that point: the complicity of Nazi scientists in the Holocaust, and that of American scientists in the deaths of the inhabitants of Hiroshima and Nagasaki. Reflecting simply on the latter, the head of the Manhattan Project, J. Robert Oppenheimer, famously remarked, "Physicists have known sin."

Of the many bioterrorism scenarios about which the American public is regularly warned, nobody seems to question the ready availability of scientists appropriately trained and willing to develop a bioweaponry program for the organization of choice. Does it not stand to reason that, if there are always enough appropriately trained scientists to pose such a virulent and ever-present threat, then perhaps the wall separating science and values isn't really so effective, and we ought to consider tearing it down and rebuilding it differently?

CULTURAL RELATIVISM

Popper's student Paul Feyerabend drew inspiration from Kuhn's work and followed Kuhn in using history and anthropology to develop his philosophical ideas about what science is. His 1975 book *Against Method* is a witty and over-the-top attack on the very basis of the Popperian philosophy of science—that there even exists a single formal method demarcating science from other approaches to and types of human knowledge.

Science is a diverse set of activities in one place at one time, after all; geologists generally don't do experiments, paleontologists commonly aren't very quantitative, astronomers certainly don't do fieldwork, biologists can rarely predict the future states of the systems they study, and mathematicians often don't even work with data. Is it conceivable that there actually could be a single method to describe "science" and yet still exclude astrology, astral projection, and Ouija boards—or is that very goal simply an illusion?

The situation gets even more complicated once we throw some time depth into it. The word *scientist* came into use only in the mid-1800s,[15] so how could it meaningfully be applied to someone like that alchemist Isaac Newton? If indeed what science is, and what scientists are, can be inconsistent across scientific fields and different for each generation, how can we specify a scientific method at all?

And how different really is the scientist reading the spectrophotometer printout from the shaman reading the entrails of a chicken? Or from the detective examining a crime scene? All are trying to decode—to understand and to render meaningful, within a particular context of values and assumptions—a set of data about the world and translate it into a proper course of action. On what basis should we privilege the scientist and the printout? Is the scientist that much more likely to be right? After all, we have already seen that most scientific ideas that have ever been held have been false—that was implicit in Popper's falsifiability criterion. And certainly all spectrophotometer printouts are not equal; there are decisions to be made about the quality of the sample analyzed and the proper calibration of the machine, any of which could explain away a printout whose information you (subjectively) do not quite trust.

So where does that leave science? Through a lens of cultural relativism, it becomes another way of producing knowledge. Not necessarily *merely* another way of producing knowledge, however; knowledge production is crucial to all manner of human existence, after all. But what should be clear is that people are always trying to make sense of things—and to the extent that science is a way of making sense of things, it can be usefully studied as one way of doing so, in contrast to other ways of doing so.

While this should not seem terribly threatening, it nevertheless proved to be surprisingly threatening to one segment of the scientific community[16]—the segment that had grown accustomed to having its authority on virtually all matters stand without scrutiny. Presumably this was because such relativistic approaches to knowledge contain an implicit repudiation of science as a source of unquestioned truth about the world. They certainly highlight the role of science as a cultural authority.

The problem is that science is a source of modern authority, yet also proud of its ancestry as skeptical of received wisdom; indeed, the structure of modern scientific practice is built in part on the idea that "organized skepticism" is intrinsic to its nature, as we see below. So, the received wisdom that tells us we should question the received wisdom constitutes a paradox. Does the skepticism central to scientific thought leave space to be skeptical of science itself? Or does the skepticism extend only to nonscientific alternatives?—an answer that seems both inconsistent and crassly self-interested.

The way out of the paradox is to see science culturally, with multiple simultaneous roles in society—notably, in this case, as both a source of cultural authority and a narrative of how the universe works. The cultural authority of science can hardly be disputed. Not too long ago, it was a standard of the advertising industry to have doctors or researchers vouch for their products, even if the products weren't good for you—like cigarettes. The advertisers knew their business; the apparent opinion of a scientist was simply more persuasive than the apparent opinion of a bus driver or an electrician. Indeed, this attitude was taken a famously absurd step further in a commercial that featured a man with a stethoscope saying, "I'm not really a doctor, but I play one on television."

But with authority, obviously, comes responsibility.

And with responsibility comes the self-interested desire not to be held responsible.

Thus, one of the paradoxes of contemporary science: we should heed science but not examine its track record too closely. If we do that, after all, we may discover that we shouldn't have heeded science before and maybe should think twice about it now. Those scientists and doctors who said smoking was good for you didn't come out of nowhere. They were available enough at a price, or maybe they just couldn't distinguish their own cultural ideas and values from their scientific judgments.

EUGENICS AND ITS DENIAL

At the turn of the twentieth century, the United States had no income tax and consequently not much in the way of federally supported social services. Demographically, rural life was being supplanted by city life, governed by the clock, based on wage labor, and increasingly populated by new, poor immigrants. Throughout the world, despots and aristocrats were slowly giving way to tycoons and monopolists.

In America, these developments led to a series of questions being posed for the first time: Why are there economic classes? Are the "have-nots" really less deserving than the "haves"? While popular socialist movements in Europe a generation earlier answered those questions in terms of fundamental human equality and against the naturalness of large differences in wealth and power, a different answer resonated among the educated classes in early twentieth-century America. The "haves," it seemed, were innately better endowed than the "have-nots," and the latter, swamping our cities in dangerous and unhealthy urban slums, were also prolific breeders. If you combine the empirical observation that the poor are outbreeding the rich with the folk prejudice that the poor are innately inferior to the rich, you come out with a prognosis of disaster—not just for the United States but for all civilization and, indeed, for the entire species.

The argument could easily be framed in scientific terms. From a Darwinian perspective, the fact that the pathetic poor nevertheless outbreed the robust rich makes the poor ipso facto more "fit"—as that word is understood biologically, in terms of reproduction. That would obviously be a subversion of the natural order, and in need of remedy. From a Mendelian perspective, one could see stupidity, illiteracy, poverty, lawlessness, and general backwardness as an alternative state to being a jolly good fellow, as wrinkled peas had been found to differ from round peas. And thus did the American geneticist Charles Davenport develop a scientific theory to account for social differences, economic stratification, and political dominance: Poor people had an allele for "feeblemindedness" that inhibited their mental development. Thus, the oppressed, exploited, or merely impoverished actually deserved what little they had; the present and future of the nation, the race, and the species lay with the good, hardy, northern European Protestant stock that made this country great.

Davenport published his synthesis in 1911 as *Heredity in Relation to Eugenics*. A New York naturalist named Madison Grant expanded on it in his 1916 best seller *The Passing of the Great Race*. Here, Grant set forward a two-pronged scientific attack upon the social problems in America, which, he argued, were racial. The racial difference highlighted here was not European versus African, however, but northern European versus southern European. Like Davenport and the paleontologist Henry Fairfield Osborn (president of the American Museum of Natural History, which Grant had helped to found), Grant focused on the racial qualities of the new waves of prolific immigrants, and on the impending racial disaster they brought. The solution, argued Grant, was to restrict the immigration of Italians and Jews, and to sterilize the feebleminded poor who were already here.

Grant received fan mail for *The Passing of the Great Race* from political figures as diverse as Theodore Roosevelt and Adolf Hitler. Scientists were no less enthusiastic. The MIT geneticist Frederick Adams Woods lauded the book in the pages of *Science,* the leading scientific journal in America. A few years later, in the *Journal of Heredity,* Woods defended the book again, with the argument that the bulk of the critical reviews had come from people of southern European ancestry, who would naturally not be disposed to its conclusions.[17]

By 1926, these scientists and activists had coalesced into the American Eugenics Society, Inc., with Charles Davenport and Madison Grant on its board of directors along with a Yale economist, a Chicago judge, and two other geneticists. Moreover, the AES boasted an advisory board that included nearly every biologist of note in the nation. The sole American geneticist who resisted any association with the AES was the fruit fly geneticist Thomas Hunt Morgan, who worked in the same building at Columbia as the great anthropologist Franz Boas. Boas had denounced Madison Grant's work immediately, but other biologists and geneticists either agreed with Grant or had no interest in alienating him and his powerful supporters in the New York science community, Henry Fairfield Osborn and Charles Davenport. Some, like the physical anthropologist Aleš Hrdlička, were disturbed about serving below one such as Grant on the AES advisory board and its committees, but not disturbed enough to resign. Others, like Harvard's physical anthropologist Earnest Hooton, might publicly distance themselves from extremist racism but privately sympathize—as Hooton wrote to Grant upon receiving a copy of one of his later books: "I don't expect that I shall agree with you at every point, but you are probably aware that I have a basic sympathy for you in your opposition to the flooding of this country with alien scum."[18]

The Johns Hopkins bacteriologist Herbert Spencer Jennings was the first to break ranks with the organized eugenicists as early as 1924, but that was merely a quiet resignation over their massaging of statistics to show Congress that immigrants from southeast Europe were more prone to crime than those from northwest Europe. The real critiques were coming not from scientists, and especially not from biologists and geneticists, but rather from other sources: civil libertarians (like the lawyer Clarence Darrow), lay journalists (like Walter Lippmann and H. L. Mencken), more religious-minded thinkers (like G. K. Chesterton), and the "soft scientists" like Boas. In fact, the eugenicists would readily dismiss their critics as anti-science, indeed even as anti-evolution.

The first geneticist to denounce the eugenics movement in a public forum was another Johns Hopkins biologist, Raymond Pearl. Pearl published his critique in a literary magazine, the *American Mercury*, and the public defection of a prominent biologist from the eugenics ranks was noteworthy enough to be carried by the news services. As it turns out, there were good reasons to fear the wrath of the scientific authorities: shortly after the essay came out, Harvard curtly withdrew its offer of appointment to Pearl, who remained at Johns Hopkins.[19]

The 1929 stock market crash and ensuing Depression ended much of the interest in a eugenics movement in the United States. Immigration of Jews and Italians had indeed been curtailed by Congress (1924), and the Supreme Court upheld the right of a state to sterilize its poor people against their will (1927). And yet the fact remained that poverty was a poor indicator of the quality of one's genetic endowment. The geneticist Hermann Muller observed "the dominance of economics over eugenics" in 1932, and the British biologist Lancelot Hogben noted that the ruling families of Europe were all beset with hemophilia but somehow nobody was talking about sterilizing or killing them on behalf of their debilitating genetic endowment.[20] The solution to social problems was actually not biological at all, regardless of what the most prominent biologists had been saying for decades.

Worse still, the Germans had developed a militaristic state in which eugenics figured prominently, taking ideas that had been developed in America and implementing them in their national policies. Wishing to give credit where credit is due, they acknowledged their debt to the American geneticist Harry Laughlin—Charles Davenport's amanuensis and drafter of the model sterilization laws in effect in many American states and in the Third Reich—by awarding him an honorary doctorate from Heidelberg University in 1936. But by then the Nazi policies were sufficiently scandalous that Laughlin was discouraged from accepting the honor in person and had to pick it up instead at the German Embassy.[21]

What makes this story interesting, however, is what happened after the war. Human genetics in the United States was forced to reinvent itself. Medical genetic issues replaced social genetic issues as the focus, and a British physician named Archibald Garrod was discovered as substitute founder of the field, to replace Charles Davenport. Eugenics itself was redefined to refer to optional (rather than coercive) reproductive possibilities, focused on medical (rather than social) pathologies, with an emphasis on the well-being of the family (rather than the nation or race). Whatever had been going on in eugenics in the 1920s was redefined as pseudoscientific; Madison Grant, for all the toleration (if not outright acceptance) by the biology community, was relegated to the sidelines. As one influential textbook put it, "Every sincere believer in the development of eugenics would do well to refer from time to time to such biased presentations of the problem as Grant's *Passing of the Great Race,* as a reminder of the extremes to which so-called 'eugenicists' of other days have gone and the pitfalls to be avoided."[22]

But of course, one did not have to venture as far afield as Madison Grant to see it. A major genetics textbook of 1925 had told students, "It is to be feared that even under the most favorable surroundings there would still be a great many individuals who are always on the border of self-supporting existence and whose contribution to society is so small that the elimination of their stock would be beneficial."[23] A few years later, one of the authors would be dean of Yale's graduate school, the other would be an outspoken critic of Nazi genetics, and that passage—indeed, the entire chapter—would be stricken from subsequent editions of the book.

The point is that until the Great Depression eugenics was neither unscientific nor even scientifically marginal—it was mainstream. It was so mainstream that, if you criticized it, you were beaten over the head with Darwin and Mendel! In retrospect, and significantly *only* in retrospect, we can see that the scientific community had inscribed its class, economic, and social interests upon its science; in a particular cultural context (without an idea of universal human rights or a major government role in social programs) and political context (totalitarianism), the scientific community not only rationalized the genocidal practice but was to some extent complicit.

But that was then, and this is now, isn't it?

Now, our purpose is not to impugn science—paranoia notwithstanding—but rather to use the lens of history to examine another feature of science, the ways in which it intersects with political power. Like any endeavor, science requires political patronage to thrive; Galileo and Bacon knew that back in the seventeenth century. But that patronage inevitably comes at a cost—namely, the freedom to question the political power that keeps science in place.

Beyond simply acknowledging the folly of biting the hand that feeds you, we need to understand that the modern political state and modern science grew up together; that is to say, they coevolved. And part of that coevolution involved cloaking the political nature of science.

The French philosopher-historian Michel Foucault called attention to the relationship between science and the modern state. Science, argued Foucault, became the means by which the state has increasingly gained power over our bodies, by becoming the official arbiter of what may be considered "normal" and "abnormal." The more narrowly science defines what is normal, the more the state controls us. Being diagnosed as abnormal—ranging at various times and places from being a witch, to being a homosexual, to being feebleminded, to having polio, or phthisis, or attention deficit disorder, or bulimia, or Gulf War syndrome, or being possessed by a demon—dictates some kind of diagnosis. Moreover, it dictates a social reaction and intervention. And most important, it determines the range of what is considered acceptable or tolerable and can thus pass without notice or comment, versus what we need specifically to do to certain kinds of people, or for them, or what they must do.

This is, to some extent, self-evident. Anyone conversant with the literature of physical anthropology can easily come up with dozens of old scientific papers "proving" that the brains of black people are somehow worse than those of white people. The measure of brain quality was often the size of the skull, or its shape. Of course, those papers are now dismissed as wrong.

Yet, at the time, they were widely considered "right." Was there a mass delusion of some sort? How could science have been so utterly incapable of judging the intellectual merit of those works? (Especially when it insists so strongly that it can indeed judge the intellectual merit of beliefs and practices that challenge its authority, such as creationism, acupuncture, or ESP?)

The answer is that science always has been, and always will be, inextricable from the cultural matrix of power, prestige, and politics that is the source of its cultural authority. As such, science will always be in an important position to defend the status quo—whether it is in the racial hygiene of Hitler, the Lamarckian genetics of Stalin, or the Anglo-American social Darwinism of the late nineteenth century. That doesn't make it bad—any more than the discovery of antibiotics and microwaves makes science good. But it also doesn't make science value neutral; rather, it makes science strongly value laden, and in ways that merit detailed examination.[24] For if we acknowledge that science is not good or bad or value neutral but rather *both* good and bad, then science suddenly assumes the burden of having to tell them apart and of having to side with the good. That is, after all, what the Garden of Eden story is actually about—the burden imposed by being human and by having the ability to tell good from bad, and the consequent responsibility to choose the former and to eschew the latter.

This is the view of science as a golem, put forth by Trevor Pinch and Harry Collins.[25] The golem was a creature of Jewish mythology, brought to life magically but lacking moral distinction and thus not fully human. It possessed great power, but since it could not tell good from evil its effect was ultimately tragic. Science, they say, is like a golem—both good and evil, but needing the ability to tell them apart to be ultimately useful. Science is a human endeavor, and thus it cannot be devoid of morality, responsibility, meaning, value, or self interest. The opposite idea, that science transcends the values, interests, or politics of its practitioners, is largely a self-interested image developed in the twentieth century.

NSF AND CUDOS

It is hard to imagine science without the National Science Foundation and its $6 billion dollar budget, but many people are old enough to remember those times. After all, NSF itself is the product of a mid-century vision of science, a radical one in which science would be subsidized by the federal government. And that massive subsidy would come with virtually no strings attached and hardly any oversight or accountability.

If that sounds almost too good to be true, it is. The very idea dates only to a 1945 memo by President Roosevelt's science advisor, Vannevar Bush, titled "Science: The Endless Frontier." Bush proposed the establishment of a large federal money pot to support peacetime basic scientific research, which had proven so valuable during the war. Bush called for a massive peacetime investment in real science, transcending politics, ideologies, and human frailties. The benefits would be spectacular.

Science, wrote Bush, is "one essential key to our security as a nation, to our better health, to more jobs, to a higher standard of living, and to our cultural progress." Transcending politics, fiddlesticks!—science is tied from the get-go to the nation, its stability, its economic vigor, and cultural progress (whatever that means, coming from an engineer). Far from representing this as a continuation of traditional scientific practice, Bush saw it as a break with the past, in which "the [federal] Government should accept new responsibilities for promoting the creation of new scientific knowledge and the development of scientific talent in our youth."[26]

This new investment of public capital would be accompanied by "complete independence and freedom for the nature, scope, and methodology of research," although the agency itself would answer to the executive and legislative branches of the government. In other words, there is a constant unarticulated tension between scientific freedom and the perceived needs of the state that is funding the scientific endeavor. Suppose scientific research began to go against the state's perceived self-interest? Is there any reason to think science ought to be, or could be, freer of state control than any other part of the state-funded bureaucracy?

Or, put another way, is it not anti-democratic to see science as free of the input of the people? Shouldn't citizens and voters have a voice in how their taxes are spent?[27]

And if you wouldn't imagine giving the Pentagon or the CIA a blank check and total freedom about using it, because you harbor doubts that they would spend the money wisely, why would you trust a bunch of egghead scientists to be any more honest, trustworthy, or responsible?

The reason we trust scientists more than generals or agents involves a carefully cultivated image of science as being precisely the opposite of what it is—commonly politicized, biased, and self-interested—in short, a human activity. In large part, the mid-century vision of "Science: The Endless Frontier" was broadly justified by another mid-century vision about science, the idea that science is fundamentally different from other human activities because of its overarching behavioral norms. These norms, argued the influential Columbia University sociologist Robert K. Merton, governed scientific research.

Merton's norms of science came to be known by their acronym, CUDOS, which ironically evokes the personal glory ("kudos") that is supposed to be stifled in the scientific ethos. CUDOS encapsulates the four qualities science is supposed to have, those that make it a different sort of activity.[28]

First, *communalism.* Lengthened from *communism,* which had acquired an unfashionable connotation and was originally the second on the list, this trait emphasizes the sharing of scientific research. Scientists work together and communicate freely among themselves.

Second, *universality.* Science applies everywhere, and consequently the conclusions it yields are applicable all over the world, regardless of language or culture. It thus transcends the familiar divisive features of the modern world.

Next, *disinterestedness.* Science pursues knowledge with an open mind and thus goes wherever the truth leads, regardless of where that may be. Thus, science transcends morality; it is neither good nor bad—applications are good or bad.

Finally, *organized skepticism.* Science succeeds because the burden of proof or evidence is so high that misinformation is constantly filtered out. Thus, false science is far more difficult to sustain than false knowledge of other kinds.

If that really is science, then it would complement Vannevar Bush's plea for government-funded science with minimal oversight, by ensuring that science could and would function as a self-policing, honest, open-minded activity, meriting the faith and money of the American public. In fact, however, there was fractious debate over the most fundamental issues of whether a national science foundation should serve the needs of the nation first or of science first; of how closely it should be overseen; and, most important, of the proper relationship between a scientist and the financial interests of patents.[29]

Another source of contention was whether social science should be included.[30] Vannevar Bush, an engineer and a conservative Republican, in fact opposed the inclusion of social science under the banner of a national science foundation. Not only did social scientists not produce much of financial value, but their work was redolent of politics, ideology, and all too commonly left-wing "social engineering" plans like those of the New Deal-ers.

Ultimately social science came to be supported by the National Science Foundation, whose own genesis nevertheless took five years and several failed bills after its call in "Science: The Endless Frontier." Bush was right to fear social science; it would later come to analyze science itself, to show the disjunctions between science as it is represented and science as it is.

Indeed, Robert Merton changed his own ideas of what the CUDOS signified. Initially, in 1942 they were intended to be a description of *how science works* and, more specifically, why a totalitarian state would ultimately be unsuccessful in subverting science to its own ends. Merton was saying that science is by its very nature democratic and thus is more compatible with a democratic political system than with a totalitarian one. A few decades later, however, Merton's CUDOS were seen to reflect not necessarily what scientists actually do, but *what they aspire for.* In other words, they now explicitly reflected the social norms of science—but, as any social scientist knows, the norms of a society are commonly at odds with what an ethnographer actually can observe.

FOUR Science as Practice

The sign by the door admonishes students, "Big Blabs Sink Labs." Is this a top-secret government facility beneath Area 51, studying the aliens whose spaceship crashed in Roswell decades ago and needing to keep a lid on it for fear of triggering mass panic?

Nope, it's a regular laboratory in a regular science department in a regular university, doing respected but largely uncontroversial work on human genetics. But the professor knows that novelty (or the perception of it) is the currency of continued success in science, measured in grants and publications. And it is not in his interests to have his friends, who are also his competitors, know too much about what his students are working on. If, as Thomas Kuhn argued, much of science is "turning the crank," then a small laboratory with a good idea is likely to get scooped if a big lab with technology and labor to spare gets wind of the small lab's idea and starts devoting its own space and students to it.

You can't fault someone for acting in their own interest, and protecting their investment, can you?

Well, only if you believe that science is really governed by the principles of CUDOS: communalism, universalism, disinterestedness, and organized skepticism.

The distinction between norms and practice—between what people think they're supposed to do and what they actually do—is a crucial one. Both are real, and both are culture specific, but they are complementary and most certainly not synonymous. In the 1960s, British ethnographers were roiled over their understandings of kinship. Students of African kinship, led by Meyer Fortes, understood their subject matter by reference to the rules they had carefully catalogued and had made some sense of. Students of Asian and Oceanic kinship, led by Edmund Leach, argued that the systematic rules of descent were *less* important for understanding other peoples' ideas of kinship than were the elaboration of those rules on the ground, as it were, and the fictions people contrived to make their actual marriage patterns fit their rules. In other words, descent systems were ways of making sense of the social world, but people still had to make sense of the sense-making system. Somehow they reconciled what they wanted to do, and what they actually did, with what they were supposed to do.

On the one hand, they have formal rules governing their behavior. On the other hand, they find ways to get around them and to do what they need to do.

Scientists are similar. To understand their behavior, you need to observe it as an ethnographer and analyze it as networks of ideas, expectations, obligations, perceptions, and self-interests. That scientists happen to be involved in the process of knowledge production is what provides a context or frame, as one might study detectives or priests. The kind of knowledge is different, but the quest is similar. The assumptions vary, but the frameworks are comparable: examine things, learn something from them, communicate its significance to others, draw a paycheck, begin cycle again.

Examine things. This is of course an arbitrary starting point, since an education in what counts as data and what may be worthy of your attention is required prior to any investigation. The data may be a DNA sequence printout, or the entrails of a chicken, or the scene of a crime; but the scientist, like the shaman and gumshoe, brings particular expertise to bear on a problem.

Learn something from them. The expert applies the knowledge somehow—depending upon the kind of knowledge and the kind of evidence. In science, this process is so unconscious and second nature that it requires another expert—a philosopher—to explain it. Cleverly making sense of the evidence from a crime scene has been highly esteemed from Sherlock Holmes to *CSI.* More broadly and anciently esteemed is the ability to make a wise decision based on the available evidence—from the Biblical patriarch Joseph through King Solomon and beyond. And of the greatest breadth, understanding how people apply local knowledge or put it into practice, is the stock in trade of the cultural anthropologist.

Communicate its significance to others. There are always appropriate channels for broadcasting particular kinds of knowledge. Language itself has quirks, for example, rooted in its symbolic and metaphorical nature; and its subset, writing, has particular conventions—tabooed words, linearity and direction, stylistic forms, abbreviations, jargon, and the like. Moreover, there are rules governing when, where, and how to communicate knowledge. Should you be in a trance, so that it seems as if the words are coming not from you but from some higher authority, with you merely its vessel or medium? Should you write in the passive voice ("The temperature was raised," as opposed to "I raised the temperature"), so that it seems as if the conclusion is independent of your own participation, with you merely its vessel or medium?

Draw a paycheck. It's not easy to be a priest or shaman. It takes years of hard work, apprenticeship, and deprivation. But if you apply yourself, move in the right circles, and get a couple of lucky breaks, you might be able to move into a position of influence or respect. Of course it won't be a nine-to-five job; it will be 24/7; it will be for life. And if you take it seriously, you're going to be conscious of preserving and protecting the authority you've earned. It's not just your own interests at stake but the interests and credibility of the very system that gave you that authority. You are in it; it legitimates and empowers you. Questioning the basis of that system of knowledge, its epistemic foundation, is almost literally unthinkable.

THE RISE AND FALL OF POSITIVISM

Why do we have such widespread confusion between what scientists are supposed to do and what they actually do, and such reluctance to see scientists incorporated within the more general realm of what groups of people do and think? And why is it therefore perceived as so threatening either to examine the relationship between scientific norms and practice or to examine scientific activity within the greater context of human activity?

Such examinations can be threatening only if we proceed from the assumption that science is some other kind of activity—different from the ordinary run of human thought and deed, in which any difference between what you're supposed to do and what you actually do cannot be systemic but, rather, can only be pathological. In this view, science is generally more of a superhuman activity, carried out by people smarter and nobler than you. And while it is hard to find anyone who will say it in quite those words, an assumption like that underlies some popular ideas of science.

And it is not difficult to see why scientists themselves might generally not rush to disabuse people of the idea, because it is a flattering portrait.

It is, alas, wrong.

To understand that idea of science, though, we have to return to the late nineteenth century, when universities were secularizing, Darwinism was grounding human existence in nature rather than in divinity, and technology (as railroads and electricity) seemed to promise a future that would be better than the past. Obviously science was very much at the heart of all these processes, and they seemed to project optimism for the fate of the human species. Early science fiction, from writers like Jules Verne and H. G. Wells, exploited this connection of science and futurism (with the exception of the occasional power-hungry madman, who controls technology for his own selfish ends).

The philosophy of positivism, popular in the late nineteenth century, complemented the optimistic view of a future world run scientifically. Building on the ideas of progress that suffused the writings of the Enlightenment, August Comte (1798–1857) developed a universal theory of history that held science to be the cumulative negation of earlier misguided belief systems. Not surprisingly, he regarded himself as a sort of pope, since the real pope was not a scientist and science was clearly destined to supplant religion.

By the late decades of the century, Lewis Henry Morgan elaborated in *Ancient Society* (1877) that society progressed through three universal stages of social organization: Savagery, Barbarism, and Civilization. James Frazer, in *The Golden Bough* (1890), saw the progress of human thought as consisting of three stages: Magic or Superstition, Religion, and Science.

Science, associated with technology and civilization, was clearly the way of the future and would lead us out of the backwardness and darkness of the past and into a new era of prosperity, happiness, peace, and wisdom. Other contemporary scholars, such as Edward Burnett Tylor and Karl Marx, saw progress in human history, although not necessarily coming in three discrete stages. But they all saw science as the engine of a glorious future that would eclipse the ignorance and irrationality of the past.

World War I, however, necessitated some reconceptualizing of these ideas. Poison gas and aerial bombing could quite reasonably be seen as the products of science and technology, but ones that were making people's lives decidedly shorter and worse. Being a resident or soldier in Ypres was no longer self-evidently superior to being a savage in the jungles of Borneo—where you did not have to worry about artillery fire or mustard gas, and where being mauled by a crocodile or bitten by snake didn't sound quite so much worse.

Of course, there had always been contrary voices against the Enlightenment ideas of universal progress. Jean-Jacques Rousseau had romanticized and glorified the life of the "noble savage" (a term first popularized in a 1672 play by John Dryden) and suggested that the condition of modern life represented not so much progress as decadence. But now, in 1918, if science was the engine of progress, and scientific progress was making wars broader in scope, easier to start, and more horrifying in carnage than ever before, then maybe we needed to see human history in some other fashion than linear.

The line, though, was a powerful metaphor. Easy to draw, easy to interpret, and easy to apply, you could see it everywhere as a guiding principle.[1] Not simply in human intellectual history, as in Comte's positivism—but everywhere. For example, in the study of life's diversity, or natural history, the linearity of the Great Chain of Being dominated scholarly thought until the mid-eighteenth century. This was the idea that all species could be ranked along a single line, in order of their similarity to humans, who sat at the top. In European social and political organization, hereditary aristocracies had been at the center of the feudal system, which featured a line of nobility stretching from the peasant or serf through knights, dukes, earls, barons, princes, and, finally, the king at the top. There were many local variations on these themes, and certainly these ideas were not perfectly transposable; the linearity of the Great Chain of Being was generally one species wide all the way up, while the political line was more like a Christmas tree, tapering as it rose.

Nevertheless, the two metaphors shared the ideas that everything had a single ordained place; that some places were better than others; and that changing places was tantamount to a subversion of nature and thus nearly impossible, and certainly not to be encouraged.

By the mid-eighteenth century, however, the Great Chain of Being was being supplanted by the idea that species should *not* be ranked according to their similarity to humans but, rather, should be grouped according to their similarities to each other. This was one of the signal contributions of the great Swedish naturalist Carl Linnaeus, who published under the latinized name Carolus and was later ennobled as Carl von Linné. By the early nineteenth century, Georges Cuvier (himself a baron, at the end of his life) would argue persuasively that a Great Chain could not exist, because there were four kinds of animals and they were so fundamentally different from one another as to defy linear ranking: radially symmetrical animals, jointed or segmented animals, shelled animals, and backboned animals.

In parallel, revolutions in the United States and France (and later, in various ways in other countries) sought to replace the linear political hierarchy with a system in which all people (or, at least, all white men) would have equal rights under the law as citizens. Here, again, rather than occupy a spot formally in relation to the king, all citizens would occupy positions only in relation to one another.

Of course in practice, natural historians still discussed the human species either before or after the discussion of all other species, thus preserving some elements of linearity; and full citizenship to women, nonwhites, immigrants, Jews, and others would come piecemeal, likewise preserving elements of linearity in political life. Nevertheless, in theory at least, the idea that things come naturally in linear series had been largely routed by the early twentieth century in both natural history and political science.[2]

To return, then, to the idea that human history itself is a sequence of improvements representing progress, it seemed a bit less obvious in the early twentieth century. A different guiding principle than the line might better express the relations of human societies at different times, and in different places as well. And that principle would be provided, oddly enough, by Albert Einstein.

THE RELATIVIST CRITIQUE OF CULTURE

The study and comparison of cultures were dominated by Franz Boas in the early twentieth century. In his paradigmatic book *The Mind of Primitive Man* (1911), Boas showed the difficulty of conceptualizing cultural progress in any other than a very narrow, arbitrary, and retrospective fashion. Comparing western Europeans and Mesoamericans in the ninth century would have yielded no basis for thinking that a few centuries later Europeans would colonize Mesoamerica rather than viceversa. A comparison of languages at any point in time or place yielded no basis on which to establish that one group of meaningful sounds or combination of sounds was any better than any other. Moreover, regarding culture as a local, integrated system of activities, thoughts, and meanings implied that comparisons would be particularly deceptive, since specific cultural forms would be shaped by local environments and histories. The life of a Kwakiutl was different from that of a Hopi and in turn different from a New Yorker. "Different" meant no more than that: just different, not better or worse. The Hopi could grow corn, the Kwakiutl could catch a salmon, and the New Yorker could hail a taxi, and they could be contrasted in any number of ways—but the cultural knowledge they possessed was local and functional and thus not easily comparable or rankable.

In 1921, Albert Einstein won the Nobel Prize for his theory of special relativity, expressed in the famous equation $E=mc^2$, which had little to do on the face of it with anthropology and everything to do with energy, mass, and light.[3] Its implication, however, was that time and space were experienced locally and would be different at high speeds—like that of light. In other words, measurements of something as obviously and intuitively universal as time actually depended upon the position of the observer, and therefore units of time measured under different conditions might not be strictly comparable.

To be sure, the conditions that would cause time and space to vary significantly would not be part of ordinary sensory experience; the effect required that you be sitting atop something going a lot faster than Christy Mathewson's fastball, something instead like a beam of light.[4] The consequences, however, implied something new and revelatory about the basic structure of the universe.

While that something had little to do with cultural comparisons, nevertheless the idea that the position of the observer (moving very fast) affects the measurement itself (of time), and that the difference can't be discerned by the observer because it is experienced entirely naturally and apparently objectively, had some strong resonance with what the Boasian anthropologists were trying to accomplish. After all, the idea of cultural progress was entirely scientific, and apparently objective, but it required a vantage point outside the system in question—Euro-American culture history—to see its falseness. The fallacy had been given a name by the late nineteenth-century Yale sociologist William Graham Sumner: ethnocentrism. Only when you were immersed in another culture could you see how effectively its communicative and social interactions worked, its ideas made sense, and its technology allowed people to thrive and breed.

In parallel with Linnaeus's work on species in the eighteenth century, the Boasian anthropologists argued persuasively that the proper frame of reference of cultural comparisons was to other, similar cultures, not to how much like modern urban Americans they might be. Thus, like the

Great Chain of Being, the idea of linear cultural progress was dismantled; rather than comparing cultures to a transcendent ideal, cultures would make more sense compared simply to each other. As early as 1887, Boas had written in the journal *Science*, "The main object of ethnological collections should be the dissemination of the fact that civilization is not something absolute, but that it is relative, and that our ideas and conceptions are true only so far as our civilization goes."

This "relativity" was evident in language, mythology, social relations, art forms—and by the 1930s, in the wake of Einstein's popularity, Boas's students were giving this principle a name: cultural relativity or, later, cultural relativism.[5] If it lacked the elegance of $E=mc^2$, it nevertheless served as a methodological foundation for American anthropology.

It also embodied a critique of modern culture. If the life of non-Western peoples is really not universally condemned to be "solitary, poore, nasty, brutish, and short" (in Thomas Hobbes's famous phrase from the seventeenth century), then there may be ways in which any particular native culture is better than ours—the implicit value judgment carried by cultural critiques from Jean-Jacques Rousseau in the eighteenth century through Margaret Mead's *Coming of Age in Samoa* in the modern age (1928). Wouldn't it be nice *not* to go through adolescent rebellion and sexual insecurity? Samoans didn't, said Mead. Perhaps we can learn from them. And what other things can we learn from other cultures to improve and enrich our own lives?[6]

Life in a society less technologically buffered against predators and pathogens would, it seems, be somewhat riskier in many ways than life in a middle-class suburb of Northern California. On the other hand, there are considerable risks posed by modern society, particularly in the densely populated lower economic strata, that would be unknown in "primitive" life. Drive-by shootings, industrial pollution, drug addiction, and automobile accidents are all risks of "modern" life. Likewise, there are less tangible social and psychological features that affect us adversely: for instance, vanity, alienation of the elderly, stress, and racism.

It is evident that there are elements of modern society that can be improved. The question then arises: What is the best way to identify possible means of improvement? The most obvious scientific answer is the rigorous study of other societies, the myriad other ways of being human and of interacting with conspecifics.

It may be worth noting that there are alternative wrong scientific answers—for example, the one at the heart of the eugenics movement. That particular answer involved equating social difference with poor health and thus treating social problems as if they were biomedical problems. Biomedical research, after all, is far more scientific than ethnographic research. And since a biomedical problem presumably requires a biomedical solution, it followed that people with the targeted "diseases" (criminality, stupidity, other nonnormative behaviors) would need to be "treated" with appropriate biomedical interventions (sterilization, euthanasia).[7] Consequently, people who looked to ethnography rather than to genetics could be dismissed by the eugenicists as unscientific. But the issue was never "science or no science"; rather, the issue was which one constituted the most appropriate body of scholarship to look to.

So, with the aid of hindsight, we are able to see that genetics was not the right answer to guide us in understanding and improving society. Hindsight, however, is often clearer than present sight. What is really striking about the inappropriateness of 1920s genetics to solve social problems is that the geneticists themselves were not too keen to disabuse the public of its faith in their science. Maybe they couldn't tell that the reasoning was flawed; maybe they were blinded by their common class and cultural prejudices; maybe they just didn't care as long as it glamorized their field; most likely, a bit of each. But the fact remains that self-interested geneticists in the first third of the twentieth century used their scientific status to oversell their craft. Dismissing contrarian voices, American geneticists themselves stood almost unwaveringly with the eugenics movement, until the Great Depression and the accession of the Nazis in Germany.

So large-scale investment in vasectomies, salpingectomies, and immigration restriction—rooted in absolute value judgments about the unequal inherent worth of large groups of people—became the recommendation and ultimate contribution of science in that age, and the detailed studies of diverse cultural forms and processes had to take a back seat.

The importance of cultural relativism—seeing diversity in human history rather than progress—became clearer in the years after World War II. Was not Germany in the 1930s both the most technoscientific and the most evil modern society simultaneously? Margaret Mead herself became something of a cult figure in the 1960s, with that generation's interest in, and legitimization of, non-European ethnicities. Small wonder, then, that the era began looking away from technological progress and back toward "nature"—with its "flower power," Earth Day, and the establishment of the Environmental Protection Agency.

This change reflected a more general repudiation (in the late 1960s and early 1970s) of the absolutist doctrine of progress, which had been a fundamental part of Euro-American thought for centuries. Technology had not brought us happiness and had certainly not brought happiness to the Vietnamese. To the extent that progress might be achieved in other areas, such as human rights and peaceful coexistence, it was going to have to be extracted forcibly from "the establishment" or "military-industrial complex," in whose financial interests it was to keep the world armed and angry.

The doctrine of cultural progress implicitly devalued the status, history, and identity of non-Euro-Americans. As those other statuses and histories were increasingly legitimized, concurrently the natural position of the white male at the top of the social hierarchy was being reexamined. Was that status really the culmination of human history, or was it a culturally situated inequality that required some rethinking? Even taken in a very narrow sense, it was clear that real progress in any area was something that needed to be sought out, carefully cultivated, and sometimes even had to be fought for. It looked hardly like an overwhelming engine of history anymore.

LIMITS TO PROGRESS

The intellectual area where the doctrine of progress persevered most strongly was in science. Partly this was because of the way scientists thought of history. The history of science, when presented by a scientist, could be given as essentially a time line of discoveries—which certainly gives the illusion of cumulation, linearity, and progress. But this is known to historians as "Whig history" or, more felicitously, "presentism"—that is to say, an interpretation of the past in terms of the present rather than in terms of itself. In this way, the only ideas that need be discussed are the ones that directly prefigured our modern ones and perhaps the stupid wrong ideas they replaced.

But how does it illuminate one event to say merely that it led to, or was supplanted by, another? The first event can be understood only in terms of its own past and present, for the second event still lies in the future—and the future can't cause the present. Consequently, the time-line approach to the history of science is not much more valuable than the time-line approach to history generally—perhaps up to about a junior high schooler's level.

A modern historian, by contrast, would note the ubiquity of multiple simultaneous discoveries of the same thing and look at the common circumstances operating at the time of those discoveries; would look at ideas that were accepted and ideas that were rejected, find that they don't map neatly onto retrospectively "correct" and "incorrect," and ask why; and would consequently look at channels of power and authority. In other words, the time-line view of the history of science doesn't really study history; it merely co-opts history in order to justify present-day beliefs. A historian, on the other hand, is actually interested in history—why scholars believed something at one point in time, what it meant to them, and how they came to believe something else at a different point in time.

So even if the history of science can be made to look linear and progressive, that is largely an illusion—the kind of data manipulation that scientists abjure when the data being manipulated are scientific rather than historical. This raises an even more important question: why should rightness (or more properly, similarity to modern consensus views) be the sole or principal criterion for evaluating scientific research of the past? Surely that criterion is applicable only in retrospect anyway; scientists must consequently have other criteria for evaluating work at a single point in time—so, what are they, and can we use them too?[8]

Moreover, the doctrine of progress goes contrary to the ordinary experiences of practicing scientists. There are always competing ideas at the cutting edge of a science and good reasons to adopt any of them; one doesn't see progress in the day-to-day workings of scientists. They collect data on certain problems; perhaps illuminate them, perhaps not; and see which interpretations of the data allow them to proceed more effectively. It is progress in the sense that a rat progresses through a maze, but not in the transcendent sense by which it is intended to characterize science.

The doctrine of progress also goes against the demographics of science. If science were progressing, in the sense that more and more questions are being settled, would we not be seeing fewer scientists and a smaller market for them to enter? Rather, we see more scientists produced and calls to train even more, which suggests that the number of questions to be answered is growing, not shrinking. One can talk one's way out of this apparent paradox, but it is significant just to note that the paradox is there. If science could reasonably or unproblematically be understood as progress, it would seem to imply at face value that fewer of its areas would remain open as time passed, and therefore *less* science would remain to be carried out. Rather, however, *more* science is being carried out, which in turn suggests that there is a lot more *to* science—niche specialties, market forces, hot problems, and the like—than simply increasing our knowledge of the world.

That's worth mulling over. *How can the domain of the unknown expand faster than the domain of the known?* If science exists to maximize the ratio of what is known to what is unknown, then doesn't the fact that the expansion of scientific knowledge seems to necessitate the constant employment of more and more scientists imply that the best way to maximize that ratio would be to stop science altogether? Why do we need more and more scientists if they are busy successfully plugging the holes in our knowledge? Where are all the new holes coming from?

Further, all scientists can recall periods of mania and torpor in their field. What accounts for rate variation in the collection of knowledge? If science can slow down, can it stall or stop or go in the wrong direction? Again, all scientists know examples of blind alleys in their field that misguided researchers for years. In biological anthropology, for example, *Ramapithecus* was promoted by powerful scholars in the 1960s as a human ancestor but is now thought to be more closely related to orangutans; Piltdown Man was widely regarded as a human ancestor from the 1910s into the 1950s but turned out to be a crude fraud. In molecular genetics, DNA was shown to be the genetic material in 1944, but a powerful advocate of proteins retarded the study of DNA for nearly a decade.[9]

Once again, it is possible to dismiss these as merely short-term counterexamples while maintaining that science manifests progress in the long term. But even that position can be sustained only by wearing heavy blinders that bracket science apart from all other aspects of culture and life. After all, how long do we have to wait before the progress manifests itself, and how do we recognize it as such when it arrives? One zealous advocate of the view that science represents cultural progress is the Oxford biologist Richard Dawkins, who snarkily asks, "When you actually fly to your international conference of cultural anthropologists, do you go on a magic carpet or do you go on a Boeing 747?"[10]

Let us ignore the fact that Dawkins is confusing science (ideas and behaviors) with technology (artifacts). Let us ignore further the fact that Dawkins is rejecting the scholarship of the experts (cultural anthropologists on the evaluation of cultural differences) in favor of a vulgar and commonsensical assertion (rather like the creationists, come to think of it).[11]

The answer, obviously, is that anthropologists take the jet plane, not a flying carpet. But that means only that our technology is more efficient than that of ancient Persia, not that our culture is superior. This distinction is crucial, for there are indeed cultural processes (emanating from human desires to make life easier, the ambitions of political entities for hegemony, and the necessity of economic institutions to maximize profits) that compel technology to be constantly improving and to be largely irreversible.

But, on the other hand . . . what of the genre of postapocalyptic thriller films like *Mad Max,* in which people of the not-too-distant future are reduced to a state of barbarism even if retaining chunks of technology such as motorcycles and guns? Technocultural regression or reversibility is at least imaginable to modern sensibilities; in cultural history it is also occasionally identifiable—as, for example, foragers trying food production and deciding they don't like it and returning to their previous lifeways.

But even that argument concedes too much ground, by bracketing off technology from the rest of culture, as if social anarchy (on the one hand) or simple pastoralism (think of the beginning of *Star Wars*) were possible alongside anti-gravity cruisers, death stars, and zappers. The fact is that technology is part of culture, and technology tends to improve, but the rest of culture is adapting to it, making sense of it, and incorporating it. And such adaptation, such incorporation, is not necessarily, nor even generally, recognizable as progress.

So let us return to Richard Dawkins's magic carpet. Very aesthetic. No annoying neighbors. Plenty of legroom. Perfect safety record. No security issues. No waiting for a bathroom to open.

The flying carpet fails in only one respect: it doesn't actually get you where you want to go. If that is your only criterion, then you will certainly opt for the jet plane over the tapestry. And you can compliment yourself on your speed and efficiency. But look at what you had to put up with! Was your life more enriched by the flight? Were you more comfortable, calmer, or happier by virtue of your airline experience?

Of course nobody is calling for a return to the savanna and trading in your Smith and Wesson for an Oldowan chopper. The question is simply whether we are talking about culture as a whole—the thoughts and acts that structure human life—when we argue about progress, or whether we reduce the diversities and complexities of life to simply its technological aspect. That would probably make anthropology a lot easier—by ignoring vast spheres of human existence—but wouldn't be very comprehensive or sensible.

Dawkins's critique is thus not of cultural relativism, for no competent anthropologist would invoke a contrast of the relative efficiencies of a flying carpet and a Boeing 747 to encapsulate the spectrum of cultural difference.[12] That a biologist could do so merely testifies to a central point of the anthropology of science: that it is very difficult for anyone, especially scientists, to extract themselves from the cultural milieu in which they work and identify the cultural biases and values that permeate their words.

The matter was artfully expressed in the play *Inherit the Wind,* in the context of teaching evolution. Evolution's lawyer turns to the jury and says:

> Gentlemen, progress has never been a bargain. You've got to pay for it. Sometimes I think there's a man behind a counter who says, "All right, you can have a telephone; but you'll have to give up privacy, the charm of distance. Madam, you may vote; but at a price; you lose the right to retreat behind a powder-puff or a petticoat. Mister, you may conquer the air; but the birds will lose their wonder, and the clouds will smell of gasoline."[13]

The flying carpet, we may add, is also ecofriendly.

CULTURAL PRACTICE

Like eighteenth-century Frenchmen, nineteenth-century Englishmen, and twentieth-century Germans, modern scientists may see themselves positioned at the acme of human achievement—but that vision entails some false assumptions and consequently doesn't stand up well.

First of all, it is awfully self-serving for scientists to try to bracket themselves off from moral concepts like good and evil, for that absolves them of responsibility. Indeed, it is just that responsibility that lies at the heart of the Garden of Eden story, now regarded by science as a quaint fable. But a fable has a point, a moral, that it is trying to communicate through the narrative. Indeed, that moral is why the story is being told in the first place. If science wishes to be free of the constraints of good and evil, free of the Biblical origin story, then what does it offer as its own moral? As I write these words, the headlines tell us of impending nuclear threats by North Korea and Iran, who obviously are employing plenty of scientists to help them be threatening. We can't separate science from the scientists who produce it and work within it. Consequently, it is hard to see how the production of scientific knowledge could be separable from its application, except in a highly abstract sense. Since scientists are trained in science while they exist as cultural beings, to bracket off the scientific from the moral is simply impossible in practice. And if the scope of comparison is to be the broad spectrum of science, then it stands to reason that the insecurity and evils brought about by scientific achievement balance out the benefits.

And we are not talking about whether science is good or bad, either. Science is both—a golem, by analogy to another Jewish myth.[14] The point is that there has to be an intellectual space available to discuss the bad, as well as the good, wrought by science. Yet simply acknowledging that science has a downside can be threatening to the most culturally blinkered and the thinnest skinned. We have a word for those kinds of people—zealots—and they have counterparts in people who think any discussion of the downsides of their most cherished beliefs, whether democracy, the Qu'ran, the free market, the State of Israel, or Leninism, places you in the "enemy camp."[15]

Science zealots are consequently interesting as anthropological subjects but have little of value to contribute to a discussion of the role of science in society, history, and culture. For that we need sociologists, historians, and anthropologists.

The first landmark ethnographic study of scientific practice was carried out in the laboratory of a noted French neurophysiologist. The ethnographer Bruno Latour (and sociologist Steve Woolgar) had a principal interest in how scientists distinguish scientific facts from nonfacts in their daily routines, and how they come to think that they have made a discovery. What they saw was that (as philosophers since Popper and Kuhn have appreciated) scientists have a very good idea of what they're going to find before they find it. As a consequence, if they don't find it, they look for mistakes in the execution or the machinery—which turn out to be easy to uncover. If they get what they expected, there is nothing to look for, even though there may be just as obvious a flaw in the experiment or equipment.

So how does an interesting or odd result become a discovery? The answer is not as clear-cut as we might like to think. Was it a fluke? Was the equipment operating properly? Can other interpretations be sustained? Is it worth communicating to others?

One of the interesting paradoxes that emerge from studying scientific discovery is the extent to which intuition—sharpened by years or decades of learning and working, but essentially the opposite of the normative view of science—can play a significant role in the process. Perhaps the most famous example is the work of the Nobel laureate physicist Robert Millikan, who was painstakingly examining ionized oil droplets in his laboratory at the University of Chicago in the early 1900s, to calculate the charge of an electron. Millikan performed his experiment and carried out his calculations many times and had a good idea when a specific result was "off" or had somehow failed. Consequently, his published paper reported on only about a third of the actual trials recorded in his notebooks, and the ones that he did publish are the ones with highly subjective qualifying assessments attached to them—like "beautiful" and "perfect." Rather infamously, however, Millikan misrepresented the role that his intuition—the familiarity with the equipment, the aesthetic conditions, and his assumption of what the answer was supposed to be—had played and wrote instead that he had carried out the experiment sixty times in succession and had gotten essentially the same number each time.[16]

Now there's an interesting paradox. Maintaining the illusion that there was no intuition or judgment call or aesthetics or subjectivity involved in producing the scientific result is so important that it can override the directive to tell the truth! That is not exactly dispassionate or objective, but it is *instrumental* behavior—intended to minimize your doubts about the scientific conclusions. Rightly or wrongly.

Indeed, the presentation of the scientific product itself—the paper—could be considered instrumentally deceptive as well. Immunologist Peter Medawar gave a talk on the structure of the scientific paper in 1963, three years after he was awarded the Nobel Prize in medicine. He asked starkly and rhetorically, "Is the scientific paper a fraud?" He answered himself in the affirmative, not in the sense that the scientific paper commonly or even regularly contains falsehoods, but in that "the scientific paper in its orthodox form does embody a totally mistaken conception, even a travesty, of the nature of scientific thought."[17]

A devout follower of Popper's ideas about the way in which experimental science works, recognizing that those ideas described his own research, Medawar wondered why the classic structure of the published scientific paper—introduction, materials and methods, results, discussion, conclusions—is presented as if the work were carried out in that sequence. That sequence suggests that the work being described had been carried out in an inductive framework: here is what I'm interested in, here are the techniques I used, here is what I obtained, here is what it means, here is how it affects the body of existing knowledge.

Rather, argued Medawar, scientific research proceeds quite differently, and the structure of the scientific paper is deliberately designed to conceal this difference. Thus, "the scientific paper is a fraud in the sense that it does give a totally misleading narrative of the processes of thought that go into the making of scientific discoveries."

The scientific paper does more than shoehorn the deductive aspect of scientific activity into an inductive framework for public consumption. It also is written to obscure agency—that is to say, people doing things, and thereby deserving credit or blame for them. Instead, the scientific paper is presented generally in the passive voice, as scientists have been trained to write. You cannot write, "I switched on the apparatus at 6 A.M.," but rather, "The apparatus was switched on at 6 A.M." This has the effect, of course, of keeping the focus on the apparatus rather than on the scientist.[18] A student of mine once tested the hypothesis that this sleight of hand was intentional, by turning in her chemistry notebook one week with her work all described in the active voice: I did this, I did that. She got a C, for the only time that semester. Or, rather, as she put it, "A C was obtained."

CONFLICTS OF INTERESTS

We are led inexorably to the position that testing hypotheses and finding out bits of how the universe works are interesting and significant components of science but not science itself. Science, rather, is the day-to-day activity of men and women, which somehow results in hypotheses getting tested and knowledge advancing. That makes it two additional things: cultural, since men's and women's activities (aside from automatic things like breathing and blinking) are invariably cultural; and subject to the diverse tugs, or conflicts of interests, that modern men's and women's daily lives invariably are.

The reason that science does not really proceed as it is expected to is that the people doing it have various things to weigh against the progress of science and their role in it. The nonnormative behaviors of scientists are not so much aberrations or pathologies as simply responses to one or another of the diverse interests that scientists have as scholars, citizens, workers—that is to say, as functioning cultural beings.[19]

Political Conflicts of Interest

The father of the atomic bomb, J. Robert Oppenheimer, famously quoted a sacred Hindu text when he saw what he had brought into existence: "I am become Death, the destroyer of worlds." With so much at stake, it's a good thing he was on our side.

The desire to bring a super-bomb—a destroyer of worlds—into existence is not the kind of ambition ordinary people harbor, unless they are seriously sociopathic. Equally obvious, the Manhattan Project was a good thing, because it helped the good guys win the war. Of course, the survivors of Nagasaki may have a different opinion; the point is, it's a good thing our side got it first, and not theirs.

But now different sides do have the bomb, and we live in perpetual fear of somebody using it against us. Thanks a lot, J. Robert Oppenheimer!

Oppenheimer knew what he had done and how, once unleashed, it could not be controlled—hence the line from the Bhagavad Gita. And yet, building the atomic bomb was the right thing to do, wasn't it? We couldn't very well have sat by and let somebody else do it, could we? Creating a super-bomb was act of super-patriotism that overrode the scientific concern for human welfare. (Ironically, Oppenheimer would later have his own patriotism called into question by Edward Teller, the super-duper patriotic father of the *hydrogen* bomb.)

Building weapons for your government makes you complicit in their use. There is not too much debate about it. After all, you can't very well build a super-bomb and expect that it will produce a cascade of rose petals. Its purpose is to destroy things and kill people, not much more than that. How can anyone reconcile the benign goals of science with the military demands of their country, much less with any sense of personal spirituality, as they participate in the development of weapons of mass destruction? Yet physicists working in U.S. weapons laboratories do precisely that. They are people, and they construct a harmonious sense of their lives and their labors, as all people do.[20]

The study of the intimate, and often subtle, relationships between science and governing is associated with the work of Michel Foucault, as noted in chapter 3. Developing weapons of mass destruction for your nation is rather more glaring than the science-power connections that interested Foucault. Nevertheless, the subtler forms of control over the body and the population through science can also be considered as conflicts of interest between the political entity that patronizes science and the formal production of knowledge that is idealized as science.

This conflict of interest is more overt in some fields than in others. The study of human diversity is invariably political in ways that, for example, the study of spider diversity is not. There was never any doubt that *The Bell Curve* (1994) was political; its coauthor Charles Murray was a political theorist. But there was some doubt about *Sociobiology* (1975), whose author was an entomologist and denied having a political agenda. Even if he couldn't see it, however, his biological theories about human behavior were political. For ants it may have been sociobiology, but for humans it was invariably sociopoliticobiology.

And to deny it, according to the academic critics of *Sociobiology*, was either disingenuous or unacceptably naïve.

The study of the behavior of nonhuman primates is also political. In hindsight, we can see the inscription of ideas about the family, gender relations, and even warfare upon animals who actually have no families, genders, or wars but only metaphorical extensions of them.[21] Indeed primatology and other scientific fields, like population genetics, have well-known national schools of thought. The Japanese primatological school, founded by Kinji Imanishi and Junichiro Itani, has tended to emphasize the integration of primate societies, while their Anglophone counterparts have tended to emphasize the individual competitive strategies of the animals.[22] The Japanese genetical school, founded by Motoo Kimura, has tended to focus on genetic differences that have no net effects on the organisms bearing them, making them different but not better or worse; their Anglophone counterparts have tended to emphasize the linear ranking of genes, some being better than others.

In either case, monkeys are behaving, and genes are mutating, but the meanings attributed to them are coproduced by the national tradition in which the scholar is educated. Not surprisingly, then, to criticize the assumptions of a rival national school might be tantamount to an act of war.[23]

Financial Conflicts of Interest

When science was the province of wealthy gentlemen, there was naturally rather little financial incentive to compromise the pursuit of knowledge. By the middle of the nineteenth century, German biochemistry was employing scientists in large number, and by the end of the century the companies owned by George Westinghouse and Thomas Edison were employing scientists as well. Edison himself was an entrepreneur above all else, and his greatest invention could be considered to be the modern industrial research lab.

By the middle of the twentieth century, tobacco companies had recognized the value of sponsoring their own scientific research, to show, first, that smoking was actually good for you and, subsequently, that in spite of what other scientists were saying it didn't really cause cancer. Later they would sponsor their own historians to show that they hadn't actually done it.[24]

All of that, however, was but a preface to the boom in pharmaceuticals and biotechnology that began in the late twentieth century. The spectacular rise of chemical solutions to emotional and mental difficulties, beginning with Valium (invented in 1963 and the most widely prescribed drug in the world by the mid-1970s), began to tether bioscience to corporate profits far more intimately than most scholars were willing to admit.[25] The development of DNA technologies initiated a complementary boom in biotechnology in the 1980s, with molecular geneticists quickly developing consultations and partnerships with start-up companies.

By 1992, leading science journals were having to develop conflict-of-interest policies that they had never had to worry about previously. Scientists had been publishing review articles promoting work being carried out by companies in which they were major shareholders. Was that fair? It wasn't against the rules, because there were no rules. In 2005, the National Institutes of Health, under pressure from Congress, was obliged to adopt stern regulations demanding full financial disclosure by its scientists.

Perhaps the oddest expression of the new relationships between bioscience and society came in May 1998, when the *New York Times* published an article on hopeful cancer therapies and quoted the codiscoverer of DNA—James Watson, soon to be head of the Human Genome Project—as saying that one line of research in particular would cure cancer in two years. The biotech company working on that line of research saw its stock price quadruple the next day. Watson himself quickly distanced himself from the statement, and the journalist who wrote the front-page article withdrew her million-dollar book proposal from the market. But the message had hit home: fortunes could be made or lost on the nature of the rapidly evolving connections among science journalists, their sources, and private-sector biotechnology.[26]

Indeed, that message was already familiar in some circles. When it came to balancing patents against patients, the courts were increasingly relying on the testimony of experts. But who was an expert, and what counted as expertise? Lawsuits were being cluttered with "junk science," and multi-million-dollar, precedent-setting liability judgments might well be hanging on the opinions of quacks.[27]

Personal Conflicts of Interest

Lord knows, some people are just "difficult." Perhaps personality traits like obsessiveness and competitiveness help steer their bearers into science because science tends to reward them. And it has been doing so since the days of its first great icon, Isaac Newton. Newton was famously brilliant, and infamously nutty and venal. His most famous grudges were held against Gottfried Wilhelm Leibniz (who independently invented calculus), Robert Hooke (who had similar interests and sometimes disagreed with Newton), and John Flamsteed (who had detailed star charts that Newton wanted). Newton's asexuality and mental instability tended to be glossed over by his admirers, nearly all of whom never knew him.

Perhaps the knowledge that you are smart and have successfully risen to a position of authority and respect brings out the worst—indeed, the criminal—in some people, such as the anthropologist John Buettner-Janusch, the virologist Carleton Gajdusek, or the geneticist W. French Anderson (see chapter 7).

These men are outliers from the scientific community. Perhaps they developed the feeling that, being so smart, they could get away with more—like the star quarterback whose shortcomings are overlooked by his admirers, however glaring or offensive they may be. The fact is that, to become a successful scientist, like a successful anything, requires some ambition, some ego, and some sangfroid—as well as the more obvious talent and opportunity. Add to these qualities the single-mindedness that comes with the dedication to a craft, and the constant pressure to be original (which seems to be an invitation to flout convention), and it is not too hard to see how a stereotype of the scientist as being antisocial and amoral might arise.

And yet, science progresses best when people work together and consider carefully the implications of what they do, does it not? And surely scientists, with the progress of human knowledge as their goal, would not be subject to the petty office politics that engross lesser minds, would they? You bet they would. The most famous public example of the social problems that can arise in the context of scientific research once again involves the molecular biologist James Watson.

Watson was awarded the Nobel Prize in 1962 for the discovery of the structure of DNA, along with Francis Crick. Sharing the prize with Watson and Crick was Maurice Wilkins, who headed the chemistry group at Cambridge that employed them. Not sharing the prize was Rosalind Franklin, whose data, which first demonstrated the now-famous double helix structure to Watson and Crick, were surreptitiously used by them. She had died of ovarian cancer in 1958.

Watson published his memoir, *The Double Helix*, in 1968 and was less than charitable to Franklin. Candidly, he explained to readers that "science seldom proceeds in the straightforward logical manner imagined by outsiders. Instead, its steps forward (and sometimes backward) are often very human events in which personalities and cultural traditions play major roles."[28]

That said, readers were treated unwittingly to a glimpse of what life must have been like for an alienated scientist—in this case a Jewish woman in a group of WASP men. Watson wrote that, distracted from talk of molecular biochemistry, "momentarily I wondered how she would look if she took off her glasses and did something novel with her hair." Then, somewhat later, "Suddenly Rosy came from behind the lab bench that separated us and began moving toward me. Fearing that in her hot anger she might strike me, I grabbed up the Pauling manuscript and hastily retreated to the open door." But readers at the dawn of the women's movement had no difficulty sympathizing with Franklin's rages as she was obliged to work alongside colleagues who were overeducated lecherous geeks.

Watson continued, "By then [our DNA model] had been checked out with Rosy's precise measurements. Rosy, of course, did not directly give us her data." Small wonder, then, that Rosalind Franklin became a posthumous icon of women scientists.[29] Personal relationships can control the success or failure of scientists, as with professionals everywhere. And everyone is entitled to an unthreatening workplace, where people can interact productively with you, without stealing your data or imagining you naked.

And, sadly, you sometimes have to fight to get such a workplace in science.

Class Conflicts of Interest

Scientists hold privileged positions in modern society. They are, at the very least, employed and smart. Regardless of your gifts, you don't acquire a job or smarts unless you have the opportunity to do so. Scientists, consequently, tend to come from very homogeneous backgrounds—from modern or economically developed nations, from families of at least modest means, and from ethnic backgrounds that have traditionally valued self-advancement through literacy and study. That is to say, they originate in familial circumstances with both the incentive and the means to invest in an extensive, advanced Euro-American-style education. And, of course, they tend to be men.

In other words, the scientific community constitutes a representative sample of neither the world nor even the United States.

How sensitive, then, to the concerns of the rest of the world, or to the rest of America, can we reasonably expect scientists to be? We can certainly hope that they are in touch with, and understand, the interests and lives of those who are different from them. In fact, however, scientists do not generally learn such things in the course of their training, nor is their track record particularly admirable in this regard.

Take, for example, the study of the detectable differences between black people and white people. It is at least conceivable that this study could proceed without class interests of the scientists affecting their work. And yet, in retrospect, we see that the study of the differences between black and white people was dominated at different times by advocates of slavery (Josiah Nott and George Gliddon in the mid-nineteenth century), eugenics (Earnest Hooton in the early twentieth century), and segregation (Carleton Coon in the mid-twentieth century)—all of whom saw their research as scientifically validating their politics.

Or consider the ideas that came to be known as social Darwinism, a scientific view of society in an age that was debating the merits of child labor laws, collective bargaining, the reciprocal obligations of labor and management, and the responsibility of the state to protect its citizens from abuse. The Carnegies, Mellons, Morgans, and Rockefellers were becoming almost unimaginably wealthy while the masses of people who worked for them in the mills, mines, factories, and sweatshops lived in squalor. Was such a situation fair? Or good?

Yes to both, argued the nineteenth-century Yale professor William Graham Sumner. Not only was it fair and good, it was *natural*. Progress in the history of life was driven by survival of the fittest, and so too with human society. The social hierarchy simply reflected an underlying natural hierarchy: rich people are the fittest and deserve what they have. Regulating their business practices and helping the poor would consequently be nothing less than a crime against nature.

Small wonder the industrialists and robber barons of the age loved his theories.

And they have ever since. Putting a biological spin on history—that is to say, explaining social and political facts as the outcome of natural forces—permits you to rationalize the acts of avarice and misanthropy that created and maintain the social inequalities you take for granted. Conflating differences of class with differences of genetics, as the 1994 best seller *The Bell Curve* did, simply offers an updated version of the theory.

The downside is that this subverts the scientific endeavor into a tool for the promotion of injustice. The apparatus and prestige of science are recruited merely to construct an apology for the wealthy classes and, worse still, to form a cudgel with which to bully and batter the helpless. That is neither good for the people nor good for science.

Science, we may recall, was marketed initially as a tool for the improvement of people's lives. If it serves instead to degrade them, then it stands to reason that most of us would be better off without it. For science to be worthwhile, it needs to make people's lives better, not worse.

Ideological Conflicts of Interest

We would like to think that science's conflicts with class and upbringing are largely unconscious, the products of an inability to see around or through the blinders of culture. This is what makes the study of science so valuable, for, like ethnography, it gives scholars a different perspective—one from outside the system—that allows them to examine the attribution of meaning, note the paradoxes, and pose questions of the people inside the system.

Unfortunately, sometimes conflicts of interest are less subtle. Mainstream nineteenth-century biology, for example, was called natural theology and consisted in large measure of coming to know God through

His works on earth. This is indeed what Charles Darwin learned in college. Natural theologians always found ways to infer God's wisdom and beneficence—the attributes that had to be there; they never found evidence for His stupidity or malice. Yet it would be just as easy to do. Think of acne and extinction; there's no necessity for either—they seemingly express either God's flaws as a designer or curses upon His creations. The problem isn't with God, it's with the intellectual exercise of trying to infer Him from the study of nature, which is made considerably easier if you already know what His attributes are.

Any theoretical system has more or less explicit assumptions. Those assumptions may be religious, like knowing what properties God has, so you can readily identify their expression in nature, or simply other compelling beliefs: patriotism, social justice, family. Here is an example from contemporary primatology.

How many different species of primates are there? The leading primatology texts of the 1980s tabulated about 170; the corresponding sources today enumerate about 340.[30] There are several possibilities to explain this doubling of the number of primate species. First, the primates might actually be multiplying like crazy. Second, hitherto unknown species of primates might be turning up all over the place. Neither of these, however, is true, and both seem to contravene the principal narrative of modern primatology—that primate species are in imminent danger of extinction.

The real explanation lies, however, in precisely that narrative. Since the 1980s, primatology has come to be increasingly dominated by conservation issues and funding. After all, chemists are certain that there will still be boron when they retire, while a primatologist has no such certainty about the mountain gorilla or aye-aye. Since conservation legislation to protect primates has tended to be written with the species in mind, it can be fairly easily circumvented by showing that other populations of the *same* species will not be adversely affected by whatever human endeavor is butting up against that legislation. The spirit of the legislation is restored, then, by elevating former populations to the status of species. The result is known as "taxonomic inflation" and is acknowledged outside the primate order as well.[31]

The immediate consequence of this taxonomic inflation is that it helps the primates survive. The conflict that arises between being students of primates and being advocates for primates is quickly resolved: if there are no primates, then there can be no primatology. Thus, scientific dispassion is important, but saving our study subjects is more important. The only sacrifice in this case seems to accompany any pretext of actually knowing, or caring, how many species of primates there "really" are. There are simply other issues that take precedence.

One could rationalize all this by saying that an abstract "evolutionary" concept of a primate species has been superseded by a more utilitarian "conservationist" concept. The point is, however, that the number of primate species we recognize serves as a kind of anchor for understanding the place of humans in the natural order, since our own species is one of them. And, somewhat paradoxically, the very scientists involved in obscuring evolutionary reality for the sake of the primates are also those entrusted with representing evolutionary reality to the public. But if anybody is going to obscure evolutionary realities, it should rather be their enemies, the creationists, shouldn't it?

FIVE The Problem of Creationism

Creationism is younger than Darwinism. It has to be that way, since creationism arose as a reaction against Darwinism. Prior to Darwinism, there was no need to have a word to denote it, since mysterious or miraculous origins of species constituted the universe of possibilities. It was not until the later nineteenth century, in the wake of the emerging cultural debates about religion and science, that creationism needed acknowledgment as a reactionary position or a movement and consequently required a name.[1]

Successive generations of readers in late nineteenth-century Christendom were obliged to cope with a wave of challenges to their beliefs. Charles Darwin suggested that species were not immutable but were genealogically linked. Thomas Huxley expanded upon Darwin's throwaway line toward the end of *The Origin of Species* ("Light will be thrown on the origin of man and his history") to try and throw some light on that origin and ancestry himself.

But these threats to traditional knowledge were largely indirect. It had long been known that many aspects of the Biblical narrative could simply not be taken literally; centuries of Jewish scholarship had been predicated on the idea that the Torah was to be interpreted properly, and that only a poltroon would take it at face value. What the Bible *means* needed to be forcibly extracted from what it *says*, and sometimes it says incompatible things. For Christians, Jesus himself casually voided all of his people's Biblical food prohibitions in Mark 7:19. Who's to say what other passages no longer count?

A common reading of Scripture in the early nineteenth century held that sin and death were connected, as explicitly articulated in St. Paul's Epistle to the Romans (6:23): "For the wages of sin is death." Nevertheless, it was abundantly clear that a lot of creatures lived and died, as evidenced by geology, long before there were people—for example, dinosaurs. So, too, was early philology showing that languages from Ireland to India were genealogically connected; and certainly modern European languages had been changing over the course of history. Spanish and French were clearly not created at the foot of the Tower of Babel but had arisen from a common ancestor—Latin—over the course of the previous millennium or so.

One could, of course, tweak Scripture in order to reconcile it with modern nineteenth-century thought. Perhaps a primordial language differentiation occurred miraculously at the foot of the Tower of Babel, and the cultural-historical processes that resulted in Spanish, French, Italian, and Portuguese were recent add-ons. Perhaps the wages of sin is death for people, but not for animals, who don't know right from wrong and are consequently incapable of sinning but are unfortunately still capable of dying. The important thing is the redemption from death afforded the human species by the renowned Galilean carpenter.

Darwin was a lighting rod for these ideological changes that swept our ancestors into the more recognizable modern secular state. In fact, however, he was just a drop in the bucket during the thunderstorm. The antiquity of the earth, the succession of life, and the descent of languages were all widely accepted in the decades before *The Origin of Species* was published in 1859. Darwin simply added another piece to the expanding naturalistic discourses that were dominating physics, chemistry, geology, medicine, and history.[2]

The culmination of the Scientific Revolution—its real threat to Christian society—was published in 1890 by a Scottish professor at Cambridge named James Frazer. *The Golden Bough* showed, by analyzing crude first-generation ethnographic literature and ancient classical sources, that there was nothing particularly unique and original in

Christian beliefs about Jesus. Virgin births, healings, teachings, sacrificial deaths, and redemptive resurrections were part and parcel of the ancient world's beliefs, and they could be found in diverse forms "out there" in the rest of the world, too. These were simply the symbolic motifs available to prescientific peoples contemplating the big questions in life. But if these were mythic elements, then where did that leave the Gospels?

Frazer's work stood on a par with Darwin's for many years. Like *The Origin of Species, The Golden Bough* has remained continuously in print for many decades since its original publication. But don't take my word for it; here's what the *New York Times* had to say in 1911:

> It has been said . . . that the two most important and influential English works of speculative thought which appeared in England during the last half of the nineteenth century were Darwin's "Origin of Species" and Frazer's "Golden Bough." Both showed an immense mastery of the facts bearing upon the subject and at the same time a wealth of critical ingenuity in interpreting them. Both summed up tendencies of thought which had been slowly maturing during the preceding decades, and both, it may be said, revolutionized the sphere of human knowledge with which they treated.[3]

Anthropologists subsequently abandoned (1) Frazer's methods, which came to be ridiculed as "armchair anthropology"; (2) Frazer's largely uncritical acceptance of primitive ethnographic records; (3) Frazer's theoretical premise that different cultures from diverse times and places could readily stand for ancestral or descendant forms; and (4) Frazer's belief that primitive cultures were simply ignorant permutations of modern culture. Magic and religion, to Frazer, were for people who couldn't handle science. On the other hand, his assumption that there is a fundamental continuity between the ways the "savage" and the "civilized" person think about the world—an assumption that German anthropologists were calling "the psychic unity of mankind"—is a cornerstone of ethnology. Regardless of its problems (Darwin, after all, had embraced the inheritance of acquired characters and invoked information from pigeon breeders alongside information from professors), the book demonstrated a basic fact: the Gospels were stories, and when examined as such their literal truth began to evaporate, like the literal truth of Genesis.

Again like *The Origin of Species, The Golden Bough* skirted the crucial issue (Did people evolve? Did Jesus die for our sins?) and allowed readers to take the final radical step themselves. But by the time Frazer published his popular abridgment of the work in 1922, with the most direct references to the origins of Christianity purged, the tide of rationalism and modern thought had all but swept away whatever was left of old-time religion.

Modern, urban America had writers like the *Baltimore Sun*'s H.L. Mencken, who could gleefully lampoon the rural religious folk: "If I hate any class of men in this world, it is evangelical Christians, with their bellicose stupidity, their childish belief in devils, their barbarous hoofing of all beauty, dignity and decency. But even evangelical Christians I do not hate when I see their wives."[4]

But down in the Bible Belt (a phrase Mencken coined, which has become so commonplace that we no longer need quotation marks around it), people were moving to fundamentalist Christianity in large numbers in the 1920s, as a stable agrarian life of the past increasingly gave way to the modern industrialized future. The charm of local time succumbed to standardized time zones in the 1880s, for the benefit of the railroads. Family farms were everywhere disappearing, sons and daughters left home for the big city to find a job and a new life in ever-increasing numbers, and the future was insecure. And this wasn't limited to the South. For example, in the 1920s preachers like John Roach Straton in New York and Aimee Semple McPherson in Los Angeles railed against evolution to large crowds and from the radio.

Largely an American phenomenon, fundamentalism began officially with a set of tracts, later collected in book form, that were being widely circulated by 1915. Fearing the liberalizing ideologies that were about to coalesce into the "Jazz Age," the Fundamentals instead called for Christians to resist the new ways and return to an older, more stringent, and curiously fondly remembered morality. Only then would something cosmically good transpire, evil would finally be routed from life, and things would return to their earlier pure and happy state.

In such broadly painted strokes, the call to fundamentalism is familiar as an instance of what is known among societies globally as a revitalization movement. A similar framework of ideas was used to mobilize people during the Native American Ghost Dance movement of the 1890s, the Kenyan Mau-Mau rebellion of the 1950s, and Osama bin Laden's Islam in the 1990s. Sometimes it involves giving up your possessions, sometimes it involves taking up arms, but it always involves a call to a new morality based on older, lapsed ways.

THE MONKEY TRIAL

Partly stimulated by the United States' entry into World War I, which prompted his resignation as Woodrow Wilson's secretary of state, three-time Democratic presidential candidate William Jennings Bryan found fundamentalism attractive. Bryan had been raised a devout Baptist and felt that Christian morals dictated a policy in opposition to both war and colonialism. (How times change!)

Bryan also felt undermined by Darwinism, which many scientists took as a justification for the very political stances he deemed most immoral. This version of social Darwinism was not so much about naturalizing and rationalizing the economic stratification of modern capitalism as it was about naturalizing and rationalizing the political stratification of modern colonialism. It was not hard at all to find demagogues, and even scientists, spouting off about the strong having to supplant or destroy the weak, and the progress of the human race riding on it. Bryan himself was particularly impressed by the knowledge that German military officers in World War I had been motivated by Darwinian ideas. And it hardly helped matters that Darwin's own subtitle for *The Origin of Species* was "The Preservation of Favoured Races in the Struggle for Life."

Couple that with the ideas (1) that we are not really made in the image of God after all, unless God is a sort of chimpanzee, (2) that there can be no Salvation without a Fall from Grace, but there was no Adam and hence no Fall, and (3) that the life and meaning of Christ himself have many basic elements in common with other non-Christian myths, and it makes a lot of sense that fundamentalist Christians might have some difficulty with the emerging propositions of modern science.

Bryan began to galvanize a movement against Darwinism. The paleontologist Henry Fairfield Osborn ridiculed him in the pages of the *Proceedings of the National Academy of Sciences,* describing a fossil ape molar found in Nebraska, *Hesperopithecus haroldcookii:*

> It has been suggested humorously that the animal should be named *Bryopithecus* after the most distinguished Primate which the state of Nebraska has thus far produced. It is certainly singular that this discovery is announced within six weeks of the day (March 5, 1922) that the author advised William Jennings Bryan to consult a certain passage in the Book of Job, "Speak to the earth and it shall teach thee," and it is a remarkable coincidence that the first earth to speak on this subject is the sandy earth of the Middle Pliocene Snake Creek deposits of western Nebraska.[5]

Alas, it spoke with a forked tongue, like the snake in the Garden of Eden. Osborn's colleague, William King Gregory, showed a bit later that the tooth in question actually came from a peccary.[6] There are still no apes known from the New World.

The origin of the human species would be the line drawn in the sand by Biblical literalists. To protect its children from the ostensible immorality implied by Darwinism, Tennessee passed a law in early 1925, known as the Butler Act, to prohibit the teaching of evolution in public schools. In Dayton, Tennessee, civic leaders contrived a plan to put their town on the map and revive its sagging economy. The American Civil Liberties Union had advertised its support for a test case of the new law. Dayton's elders found John Scopes, an unmarried teacher and upstanding citizen, and prevailed upon him to be arrested for violating the Butler Act, so that Dayton could be the site of just such a test case.

Things, however, quickly got out of hand. William Jennings Bryan, regarded as the greatest orator in the land, volunteered his services to the prosecution. Chicago's Clarence Darrow, regarded as the greatest trial lawyer in the land, came to assist the defense. Darrow took on the case pro bono, for the first and only time in his illustrious career.

H. L. Mencken aggressively recruited the *Baltimore Sun*'s financial support, as well as Darrow's participation. Along with Darrow, for the trial of John Scopes in July 1925, came the prominent New York lawyer Dudley Field Malone. The *Baltimore Sun* brought experts down to Tennessee, from rabbis and ministers to geologists and anthropologists—but all such testimony was rendered inadmissible by Judge John Raulston. He would stick to the narrow question of whether Scopes had violated the law, not whether the law itself was inane—which is where the defense wanted to lead the case.

Meanwhile, Mencken's searing prose for the *Baltimore Sun* was being reprinted across the country:

> The Book of Revelation has all the authority, in these theological uplands, of military orders in time of war. The people turn to it for light upon all their problems, spiritual and secular. If a text were found in it denouncing the Anti-Evolution law, then the Anti-Evolution law would become infamous overnight. But so far the exegetes who roar and snuffle in the town have found no such text. Instead they have found only blazing ratifications and reinforcements of Genesis. Darwin is the devil with seven tails and nine horns. Scopes, though he is disguised by flannel pantaloons and a Beta Theta Pi haircut, is the harlot of Babylon. Darrow is Beelzebub in person.

> The Bryan of today is not to be mistaken for the political rabble rouser of two decades ago. That earlier Bryan may have been grossly in error, but he at least kept his errors within the bounds of reason: it was still possible to follow him without yielding up all intelligence. The Bryan of today, old, disappointed and embittered, is a far different bird. He realizes at last the glories of this world are not for him, and he takes refuge, peasant-like, in religious hallucinations. They depart from sense altogether. They are not merely silly; they are downright idiotic. And, being idiotic, they appeal with irresistible force to the poor half-wits upon whom the old charlatan now preys.[7]

The *New York Times* ran a story on July 17 with the headline "Mencken Epithets Rouse Dayton's Ire" and noted that some of the town's residents were considering "taking him into an alley."

Ultimately, the defense came up with a last-ditch strategy. Unable to call scholarly witnesses to impeach the Bible, they would play on Bryan's ego to get him on the witness stand and cross-examine him as an expert on the Bible. Bryan understood that he would likewise question Darrow afterward and took the stand on the afternoon of Monday, July 20. It was so hot that day that the trial was moved outdoors. Ironically, many of the journalists, including Mencken himself, had decided that the most interesting parts of the trial were over and had left town over the weekend.

Over two hours that afternoon, Darrow relentlessly grilled Bryan on the absolute veracity of the Bible. Did Joshua make the sun stand still in the sky? Yes. Did a great flood kill all living things except those on Noah's ark? Yes, except perhaps for the fish. How did the serpent move around before being cursed to crawl on its belly? Don't know. Did Jonah really swallow the whale? Yes. (Hah! Trick question!)

The most critical testimony, however, was when Bryan volunteered that he was not as much of a literalist as he had been made out to be. How long were days described in Genesis? Of indeterminate length, not necessarily twenty-four hours. Hot and tired, and not really wanting to answer Darrow's question about the origin of rainbows, Bryan finally exclaimed, "Your Honor, I think I can shorten this testimony. The only purpose Mr. Darrow has is to slur at the Bible, but I will answer his question. I will answer his question. I will answer it all at once, and I have no objection in the world. I want the world to know that this man, who does not believe in a God, is trying to use a court in Tennessee to slur at it, and while it will require time, I am willing to take it."

To which Darrow shouted back, "I object to that! I object to your statement. I am examining you on your fool ideas that no intelligent Christian on earth believes!"

Court was immediately adjourned and Bryan's testimony stricken from the record. That Scopes was convicted, and the verdict overturned on a technicality, became largely inconsequential. The religious progressives had squared off against the backward-thinking zealots and had sent them back to their pews, licking their ideological wounds.[8]

Bryan, interestingly enough, died in his hotel room in Dayton just a few days after the trial, a sad fact that had absolutely no theological significance, as the biologist Julian Huxley later observed. On the other hand, if it had been Darrow or Mencken who had dropped dead just after the trial . . .[9]

CREATIONISM AFTER SCOPES

While evolutionary theory in the United States blossomed in the intervening decades, pedagogically it was driven underground by market forces until Sputnik reawakened public interest in science education.[10] In step, however, anti-evolutionary sentiments began emerging as well. A hydraulic engineer named Henry Morris reinvigorated it, with books purporting to explain geological features in terms of Noah's flood, with a six-thousand-year timetable for the history of the universe. He later founded the Institute for Creation Research, in 1970.

Making good use of religious networks on college campuses and the tendency of the scientific community to ignore them, the creationists established beachheads in the intellectual mainstream and adopted a new legal strategy. Having failed in their attempt to outlaw evolution, they would now argue that creationism (i.e., Biblical literalism) is an alternative scientific view and try to get them taught together.

This second creationist legal strategy resonated with American concepts of fairness and equal time. Buoyed by successes at the local level, Arkansas passed an "equal time" creation law, mandating that "public schools within this State shall give balanced treatment to creation-science and to evolution-science." With the power of a conveniently discovered relativism, the bill continued, "Creation-science is an alternative scientific model of origins and can be presented from a strictly scientific standpoint without any religious doctrine just as evolution-science can, because there are scientists who conclude that scientific data best support creation-science and because scientific evidences and inferences have been presented for creation-science."[11]

In his decision for *McLean v. Arkansas*, sent down on January 5, 1982, U.S. District Court judge William C. Overton concluded that creation-science was charlatanry, its resemblances to science entirely cosmetic. The judge opined, "It was simply and purely an effort to introduce the Biblical version of creation into the public school curricula."[12]

Scientific creationism went to the Supreme Court in 1987, and, with Justice Scalia and Chief Justice Rehnquist dissenting, a 7–2 decision again saw right through scientific creationism and ruled that Louisiana's "equal time" law "impermissibly endorses religion by advancing the religious belief that a supernatural being created humankind." On the other hand, a pessimist could be apprehensive at the prospect that two judges on the Supreme Court couldn't see through the sham—observing instead that "political activism by the religiously motivated is part of our heritage" and specifically denying that the purpose of the act was "to advance religion."[13]

As William Jennings Bryan had been the political-legal force behind trying to get evolution outlawed, the move to get evolution "balanced" against the Bible was spearheaded by a lawyer named Wendell Bird. But by the time the decision was handed down, a Berkeley law professor named Phillip Johnson was devising a third legal challenge to evolution.

Johnson argued not that creationism is scientific but that science itself is biased against other modes of thought, in particular those that invoke supernatural explanations. He encouraged evangelical Christians strategically to dispense with talk of God and instead talk of a Designer. In this way, they could cast their net beyond the Protestant Biblical literalism that motivated them and embrace people from diverse religions and denominations, who might share no common bond other than faith in a purposive universe. With the support of a Seattle-based "research and policy organization," the Discovery Institute, these ideas were introduced as a "wedge" by which to topple Darwinism (again), and indeed they enjoyed short-term success.[14]

This "intelligent-design" library was never very large and was always somewhat incoherent. Phillip Johnson's books attacking evolution never actually articulate alternative explanations or scenarios for the history of life. Trying to embrace Biblical literalists as well as "people of faith," Johnson deliberately refrains from saying how old he thinks the earth is, or what the geological column or the succession of life mean. He readily acknowledges molecular, population, and ecological genetics as describing and explaining "minor" aspects of the history of life accurately but disputes their relevance in explaining "major" changes—and he doesn't explain how you make the crucial distinction between a minor (naturalistic) from a major (miraculous) change.

Similarly, the Catholic cell biologist Michael Behe promoted the idea of "irreducible complexity"—that anything you cannot envision as having been built up stepwise must instead have come into existence all at once, and by an "intelligence," presumably divine. He sees aspects of the cell in such a fashion. On the other hand, he writes that Darwinism is not so much wrong as "incomplete" and adds, "I have no quarrel with the idea of common descent, and continue to think it explains similarities among species."[15]

The other major works of the canon are by the philosopher William Dembski, who proves mathematically that evolution cannot happen yet paradoxically seems to believe that miracles can, and by a member of Sun Myung Moon's Unification Church, the biologist Jonathan Wells. Wells, like creationists generally, has little to say in support of creationism but a lot to say against evolution—and, even then, not so much about evolutionary theory as about the moral turpitude of scientists.[16]

By cleverly pitching it at the lowest common theological denominator—are we here for a reason, or not?—the intelligent-design movement managed to make its Biblical literalist agenda interesting even to some Catholics and Jews. Once again, since intelligent design was ultimately a legal strategy intended to undermine science education in the United States, it was up to the courts to decide whether normative science was all that was to be taught in public school science classes. President George W. Bush had already lent his weight to the idea that intelligent design should be taught alongside evolution.[17] But his appointee, Judge John Jones III, who was asked to rule on the decision of a local school board in Pennsylvania to bring intelligent design to the attention of students as a scientific alternative to evolution, saw little merit in it. In *Kitzmiller v. Dover Area School District* (2005), Judge Jones saw the same duplicity as his predecessors, acknowledging that intelligent design "is a religious view, a mere re-labeling of creationism, and not a scientific theory."

The point is that intelligent design, the latest legal challenge to Darwinism, is not so much a theory as an *anti*-theory. It stands *for* nothing but simply *against* the proposition that the human species arose by natural processes from other species, and that oppositional stance seems to have broad resonance. It is a theory about the wrongness of evolution but not about the rightness of any alternative.

NATURAL THEOLOGY AND THE ARGUMENT FROM DESIGN

Discovering the regularities that govern the workings of the universe could be seen as glimpsing into God's mind. This would have been the height of blasphemy a few hundred years earlier, but by 1700 it could be taken as a respectable—indeed, reverential—goal. Isaac Newton was hailed as one who had decoded some of God's deepest secrets, and the merits of continuing in that vein were taken as self-evident.[18] Three centuries later, after sequencing the human genome with millennial technology, the geneticist Francis Collins invoked a familiar metaphor: "We have caught the first glimpse of our own instruction book, previously known only to God."[19] As noted in chapter 2, however, there is also a case—and a darn good case—to be made for *not* trying to find out what God alone knows.

But by the eighteenth century this moral had indeed been reversed. What had seemed heretical a couple of centuries earlier actually became a source of piety, stretching up unto the Human Genome Project at the end of the twentieth century. God apparently now *does* want us to know His secrets (although He still doesn't make it too easy for us).[20]

The point is that striving to find out how the universe works is not self-evidently a good thing. How light works or how gravity works is not manifestly good knowledge, as long as you are able to see when you need to and can get up when you fall down. On the other hand, learning how to be a law-abiding citizen, a responsible parent, a sympathetic companion, a persuasive speaker, a polite neighbor—are all good kinds of knowledge, but not of *scientific* knowledge. There are many kinds of knowledge, of different value in their appropriate contexts. And, as writers from Christopher Marlowe *(Doctor Faustus)* through Mary Shelley *(Frankenstein)* and up to Michael Crichton *(Jurassic Park)* have been trying to tell us, amassing knowledge without wisdom is dangerous.

The most obvious way to associate knowledge with wisdom seemed to be to link science (knowledge) to religion (morality); but this would have to be done very carefully, insofar as science is to a large extent predicated on the construction of a barrier between the natural domain of science and the human domain of aesthetics, spirituality, politics, and the like. Biology and religion were transiently yoked together over the eighteenth century by English scholars, beginning with John Ray's *The Wisdom of God Manifested in the Works of Nature* (1691), culminating in William Paley's *Natural Theology* (1802), and ending with the diverse Treatises commissioned by the Earl of Bridgewater in the 1830s. These scholars believed that the greatest goal of the study of life was to testify to the power, beneficence, and bounty of God. After all, as Creator of the Universe, God must have left His imprint upon it. So why not look for it?

Paley's famous example was drawn from medieval Christian theology and has come to be known as the "argument from design." It takes God to be the separator of Order from primordially "unformed and void" Chaos. God's stamp upon the world, then, is complexity itself. Paley meditated on a watch, whose intricacies and precise mechanisms implied the existence of a watchmaker, and concluded that intricacies and precision in nature implied the existence of a designer of nature.

This was in fact the biology that Charles Darwin studied while a student at Cambridge and Edinburgh. But, for all its piety, natural theology was increasingly difficult to sustain as the theological meaning of biological patterns in nature became more obscure. John Ray understood God's wisdom to reach expression in the Great Chain of Being, which connected all living forms to each other linearly; consequently, he pronounced extinction impossible, since it would represent the destruction of God's plan. And yet by the mid-1700s, two things were clear: extinction was for real, both in the present (as in the case of the dodo) and especially in the past; and the Great Chain of Being didn't seem to describe nature very well, for there didn't seem to be any obvious way of telling, for example, whether a bear was higher or lower than a bunny. Clearly if God's plan really were manifested in the works of creation, it was a somewhat different plan than the one that Ray saw.

Natural theology was what Darwin destroyed, the idea that the history of life—that is, the origin (and termination) of species—could best be understood as a series of purposeful miracles, and that God's purpose could be revealed by appropriate study and meditation. The origin of life itself might still be a miracle, but the number of such miracles shrank to "one or a few" in the last paragraph of *The Origin of Species*. Darwin showed that the history of life could be explained by understandable natural forces and, consequently, probably ought to be. Whatever God's plan was, it was not accessible to science, that is to say, by the conventions of reliability established since the seventeenth century. The glory and wisdom of God were nice, but they could be safely isolated from a discussion of the diversity of life, whose patterns were more satisfactorily explained by common ancestry.

In some sense Darwinism is an automatic outgrowth of rationalism: given a set of methods for thinking about the world, it stands to reason that sooner or later the explanatory power of common descent will be appreciated (as in fact it was by several other scholars in Victorian England). It is simply the theory that explains the data best, which is what science aims to produce.

Intelligent-design theory, the new version of creationism, is a return to natural theology. It is perfectly good science, but for 1820. Intelligent design is framed to conflict not so much with the Darwinian interpretation of the data on the history of life as with the intellectual methods by which data are interpreted. In other words, it is not specifically about whether we came from apes but about how we draw scientific inferences. It is about what counts as reliable and useful knowledge: is the naturalistic explanation the best, or should supernatural explanations be admitted into the arsenal of scientific tools as well?

You don't need ESP to sense the parapsychologists drooling in anticipation of legitimacy.

ARGUMENTS AGAINST INTELLIGENT DESIGN

The first problem with intelligent design is that it actually presents no alternative theory to the ones it seeks to supplant. "Young earth creationism" is the idea that the universe is either being misread to seem older or was intentionally made to look older but is really only about six thousand years old. There's little to be said about this, since it prioritizes a narrow reading of Genesis over all other sources of knowledge, encodes a basically paranoid view of nature and of the people who study it, and was consequently too stupid even for William Jennings Bryan to believe. "Old earth creationism," on the other hand, while accepting the sciences of astrophysics, geology, and archaeology, still has to explain biological patterns of similarity and distribution.[21] And if common descent and adaptive divergence aren't their explanation, then its adherents are obliged to say what is.

Intelligent design's second problem is that there is actually very little new to say about the subject. Indeed, David Hume had philosophically dismantled the ideas behind natural theology in his *Dialogues Concerning Natural Religion* (1795), and his arguments apply as readily to intelligent design more than two hundred years later.

Hume pointed out, as the ancient Greeks knew, that arguments from ultimate causes lead to infinite regresses. If God must be invoked as the cause of this world, then what is the cause of God and His world? And the cause of that cause? Sooner or later, you just have to stop and say there is an uncaused cause, and you can call that God if you want. But where you stop is arbitrary, and ultimately all you have done is construct a nest of imaginary universes by which to try to explain the only real one. You're better off not starting that game at all.

Moreover, who says the body is like a watch or, in Hume's argument, that the universe is like a house? It may be useful as an analogy or heuristic to see the universe as a house, but it is not literally so. The universe may be like a house in some ways, but since it isn't really a house it therefore doesn't necessarily have any of the specific properties of houses. Likewise, a body may be conceptualized to be like a machine with some valuable results (e.g., William Harvey's work on the circulatory system in the 1620s). But it isn't a machine and therefore doesn't necessarily have to have the same properties that machines do (like a designer).

Finally, as Hume wryly observed, "If you find any inconveniences and deformities in the building, you will always, without entering into any detail, condemn the architect." So why don't the imperfections and evil designs of nature—acne, hemorrhoids, extinction, cancer, PMS, hay fever, tapeworms, scoliosis, baldness, impacted wisdom teeth—demand a condemnation of their Creator? Could it be because natural theologians have assumed Him rather than deduced Him?

Intelligent design's third problem involves the reactive aspects of nature in confrontation both with the world itself and with the human mind. Toward the external world, living systems are homeostatic; that is, they are responsive and self-regulating. This obviously tends to produce a fit between the system and its surroundings. These homeostatic mechanisms are not conscious, but they result in a continuous process of negotiation between, for example, a species and its niche. Sometimes, however, consciousness may play a complicating role as well—for a natural object can enter into a reactive relationship not only with its surroundings but with the human mind thinking about it. If a cloud can remind Hamlet of a camel, a weasel, or a whale, that raises a question not just about how clouds acquire shapes but about the meaning of the shapes themselves.[22] Was the cloud fashioned to remind Hamlet of the animals (and to help him ridicule Polonius), or does human imagination impart the idea of design wantonly, because human minds impart meanings to things? If the latter, then what tests are available to distinguish between things that seem designed because they *seem* to have an evocative form (like clouds) and things that were indeed designed because their designer *wanted* them to have an evocative form (like Michelangelo's *Pièta*)? If there are no such tests, then the designer being inferred is as likely as not to be a figment of the observer's imagination.

The final problem is that intricacy and organization can be readily inferred even from things that were clearly not consciously designed. Snowflakes are intricate, but does anyone really think that God unleashes a horde of microscopic chiselers in constructing a blizzard? I may not know much about the physics of crystal formation, but it's got to be a better explanation than that one. And Italian dressing is highly organized: each component seems to know its place and head right for it—with the oil at the top, the vinegar in the middle, and the herbs settling at the bottom, in just a few seconds, no matter how vigorously you shake it to get its components mixed up and disorganized. Angels don't push them into place; they just have different densities and react accordingly under the influence of a gravitational field.

So if you can get intricacy and organization in contexts where the direct hand of God would be patently ridiculous, then those properties are not very reliable as proof of the existence of God. And even if God Himself did dictate the laws of cloud formation, crystal growth, and mass, the sad fact is that science is interested in what the laws say, not in where they came from.

At this point, however, some cultural relativism may be useful. Granting that scientists are indeed interested in the laws, rather than in their source, is it possible that other people might be more interested in the source than in the laws? Is the interest of science necessarily the best or only interest?

Speaking for myself, I can say in all sincerity that I think the question of God is more interesting than the question of crystal growth. I don't know whether that is because the question of God is so inherently fascinating or because crystal growth is just so damn boring.[23] Think of watching paint dry in three dimensions. What is important is to recognize that they are separate questions, which is why I would not make a good natural theologian.

David Hume, back in the eighteenth century, was already suspicious of anyone drawing too close a connection between the things created by human agency and the wonders of nature. After all, we can see the products of human thought and labor continually being improved, but we don't see nature acting similarly.

> If we survey a ship, what an exalted idea must we form of the ingenuity of the carpenter who framed so complicated, useful, and beautiful a machine? And what surprise must we feel, when we find him a stupid mechanic, who imitated others, and copied an art, which, through a long succession of ages, after multiplied trials, mistakes, corrections,

> deliberations, and controversies, had been gradually improving? Many worlds might have been botched and bungled, throughout an eternity, ere this system was struck out; much labour lost, many fruitless trials made; and a slow, but continued improvement carried on during infinite ages in the art of world-making.

This is a scenario—a series of imperfect creations, each more closely approximating the modern one—that seems to emerge from taking the analogy between man-made things and God-made things too seriously. But with (1) natural theology as the principal explanation for life's patterns, (2) animal forms revealing themselves to be more exotic and unfamiliar the further back in time paleontologists went, and (3) the history of life ridden with appearances and disappearances all the way through, Hume's extreme scenario actually anticipates some of the more conventional scientific explanations that emerged in the decades before *The Origin of Species.*

The issue is not "Does God exist?" but rather "What can we reasonably infer about God from observing nature?" If the answer is "nothing," then we cannot rely on our senses to apprehend God, and all we have to fall back upon is faith. In some religious quarters that would be enough, but not for the natural theologian.

CLARENCE DARROW'S ARC

Preparing for the Scopes trial in 1925, Clarence Darrow did his homework and actually read the textbook that students had been assigned. It was George W. Hunter's *A Civic Biology,* and it indeed gave a perfunctory description of evolutionary processes, as they were understood at the time, and an evolutionary interpretation of the history of life.

Then it went on to the evolution of people and its culmination in the most advanced form—the white, or Caucasian. A bit later, it explained to students about the innateness of intelligence and the need to sterilize the unfit. Darrow, who had spent his life opposing capital punishment for the commission of a crime, found the argument for sterilization on the grounds of one's condition at birth—as if being born could somehow be a crime—repugnant. Even worse, it seemed to be justified by Darwinism, or at least it was presented that way.

Nor did the textbook get it wrong. Geneticists themselves widely agreed that the poor were outbreeding the rich yet were genetically worse than the rich, so the government ought to do something about it. What they settled on was a program of restricting the immigration of southern and eastern Europeans (i.e., Italians and Jews) and involuntarily sterilizing the poor. This was the American eugenics movement.

What is interesting in the present context is the association between eugenics and Darwinism. In fact, the British eugenics movement was initiated by Darwin's cousin, Francis Galton—who, indeed, was responsible for the very word *eugenics*—and was later headed up by Charles Darwin's son, Leonard.

In the United States, biologists scrambled over one another to speak for both evolution and eugenics. The geneticist Raymond Pearl wrote glowingly of the First International Congress of Eugenics in 1910: "Eugenics is distinctly an applied science. Hitherto everybody except the scientist has had a chance at directing the course of human evolution. In the eugenics movement an earnest attempt is being made to show that science is the only safe guide in respect to the most fundamental of social problems."[24]

The paleontologist Henry Fairfield Osborn told the Second Congress in 1921: "The 500,000 years of human evolution, under widely different environmental conditions, have impressed certain distinctive virtues as well as faults on each race. In the matter of racial virtues, my opinion is that from biological principles there is little promise in the 'melting pot' theory."[25]

And the leading human geneticist in the United States ended his address to the Second Congress with these words:

> The human species must eventually go the way of all species of which we have a paleontological record; already there are clear signs of a wide-spread deterioration in this most complex and unstable of all animal types. A failure to be influenced by the findings of the students of eugenics or a continuance in our present fatuous belief in the potency of money to cure racial evils will hasten the end. But if there

be a serious support of research in eugenics and a willingness to be guided by clearly established facts in this field, the end of our species may long be postponed—and the race may be brought to higher levels of racial health, happiness and effectiveness.[26]

Clearly there was a pretty strong association between eugenics and Darwinism on the part of biologists on both sides of the Atlantic. It might almost seem that to be against eugenics was to be against evolution. Certainly the eugenicists did their best to promote that conjunction of ideas.

As noted in chapter 3, it was very hard to find a biologist critical of eugenics until quite late in the game. In England, the leading geneticist, William Bateson, was anti-eugenics, but largely because he hated Britain's other leading geneticist, Karl Pearson, who was an ardent eugenicist. In the United States, the bacteriologist Herbert Spencer Jennings became disillusioned after examining the data the scientific community presented to Congress in support of the 1924 immigration restriction bill. Jennings published his analysis and quietly distanced himself from the movement. Shortly thereafter, so did his colleague Raymond Pearl.

The fruit fly geneticist Thomas Hunt Morgan was the only American biologist to keep eugenics at arm's length throughout the 1920s. Nevertheless, he too refused to criticize eugenics directly. Indeed, until the late 1920s the only criticism of eugenics came from nonbiologists, who generally risked being tarred as creationists for their criticisms of eugenics. The anthropologist Franz Boas published a comprehensive critique of the ideas of the eugenics movement in 1916, but to little effect. The journalist Walter Lippmann published a series of articles criticizing the use of IQ tests in 1922. The English novelist G.K. Chesterton wrote the same year that the eugenicists "have fathers and mothers like other people; and our opinion about their fathers and mothers is worth exactly as much as their opinions about ours."[27]

For the present context, however, let us turn to the culture heroes of the Scopes trial, H.L. Mencken and Clarence Darrow. Shortly after the conclusion of the trial, Darrow wrote an article for Mencken's literary magazine, the *American Mercury*, on the quackery coming out of genetics and psychology on the inheritance of "feeblemindedness." He singled out the scientific studies that were being most widely brandished in support of sterilizing the poor, namely, family studies purporting to show that feeblemindedness is genetic (which actually documented that poverty is inherited and dictates one's life trajectory more assuredly than one's genotype does) by counting up the number of great people and loathsome people in family trees.

The most famous of these families were the Kallikaks, a pseudonym conferred upon them by the psychologist Henry H. Goddard. Through a "nameless feebleminded tavern girl," and subsequently through his lawful wife, "Martin Kallikak" sired two lines—one adorned with lawyers and doctors, the other full of bums and sluts. You can probably guess which one was which. The weakness of the argument notwithstanding (and we now know that the author lied publicly in response to the question "How did you know the ancestress was feebleminded if you didn't even know her name?"), the work would be referenced in textbooks of genetics and psychology for decades to come.

Darrow, however, had different targets: the Edwardses and the Jukeses. Eugenicists brandished these studies alongside the Kallikaks, as did high-school textbooks like the very one that John T. Scopes used.

Jonathan Edwards was the progenitor of one such "great line." A prominent fire-and-brimstone preacher, he later became president of Princeton. Among his descendants were many prominent citizens; did this not demonstrate the stable inheritance of good character? And the descendants of Max Jukes went in the opposite direction, being notable for the drunkards and reprobates among them. It hardly taxed Darrow's analytic skills to dismantle the case for inferring that prominence was passed on genetically on the basis of crude evidence like this.

Darrow was at great pains to distinguish science from charlatanry and was determined to expose the latter, even if the scientists themselves had difficulty in telling them apart from one another. A few months later, again in H.L. Mencken's *American Mercury,* Darrow took on the mainstream eugenics movement. "Amongst the schemes for remolding society," he wrote, never one to mince words, "this is the most senseless and impudent that has ever been put forward by irresponsible fanatics to plague a long-suffering race."[28]

Darrow had come to appreciate two things. First, eugenics needed to be disentangled from evolution, however reluctant biologists might be to separate them. Second, when it came to social and political issues, knowledgeable biologists were almost stunningly ignorant. And yet they brought the same arrogance to a discussion of social issues, about which they knew little, that they brought to a discussion of the origin and history of life, about which they knew much.

In the *Baltimore Sun*, Mencken himself took on the eugenicists. "I am convinced that their great cause is mainly blather," he wrote.

> The fact is that the difference between the better sort of human beings and the lesser sort, biologically speaking, is very slight. There may be, at the very top, a small class of people whose blood is preponderantly superior and distinguished, and there may be, at the bottom, another class whose blood is almost wholly debased, but both are very small. The folks between are all pretty much alike.[29]

Six months later, in May 1927, Mencken published (in the *American Mercury*) the first head-on critique of the eugenics platform by the hand of any American biologist. It was written by his friend, the Johns Hopkins biologist Raymond Pearl. Pearl's essay, "The Biology of Superiority," was the subject of news stories around the country. "Like Not Produced by Like, Biologist Says in Attacking Eugenics," declared one. "Eugenics All Wrong about Better Babies, Savant Finds," proclaimed another.[30]

Pearl's essay was newsworthy not for what it said—which had been said by others in the preceding two decades—but for who was saying it. Somehow, somewhere, a practicing, reputable biologist had finally wised up.

Good thing the Monkey Trial was over.

ANTHROPOLOGY AND CREATIONISM

At root, Darwinism simply relates the diverse adaptations possessed by species to a history of differential survival and reproduction of the organisms bearing slightly different features. The initial differences are produced by a random process we now call mutation, but those random variations are sorted by a decidedly nonrandom process in which the environment "selects" the organisms that will proliferate (much as an animal or plant breeder might select the parents of the next generation) over immensely long periods of time.

Adaptation becomes not so much a state endowed magically from the beginning of time, but a historical process. The correspondence of parts across diverse species, known as homology (such as the similarity in structure of the pigeon's wing, dog's forepaw, and human arm), becomes not so much a mysterious lapse of divine creativity, but a remnant of the extent of common ancestry shared by those species. Dogs can now be seen as wolves modified for domesticity, birds as dinosaurs modified for flight, and people as apes modified for walking and talking.

Darwin, however, quickly became bigger than Darwinism. An ancient philosopher reputedly said, "For many shall come in my name, saying, I am Christ; and shall deceive many." A contemporary and admirer of Darwin reputedly said, "Je ne suis pas Marxiste."[31] Marx denies being a French Marxist, and Jesus warns against impersonators. The message is the same: Don't be fooled! Both men were eponymous heads of movements that embodied ideologies reaching far beyond anything they had ever actually said or implied. So too with Darwin and Darwinism.

The opposition to Darwin and evolution is really not about Darwin himself, nor about the transformation of gene pools, but rather about what they signify. Because Darwin's name acquired the cultural power that comes through an association with science, modernity, and progress, it is perennially invoked in support of ideologies that have only the most tenuous connection to the study of the genealogy of life. In previous generations, it was social Darwinism and eugenics. Today, scholars interested in reifying racial categories (in defiance of known patterns of human diversity) commonly cloak their arguments in Darwinism. Taking that position for granted, some scholars proceed to act as shills for the pharmaceutical industry, promoting the construction of racialized niche markets for their products.[32] A leading spokesman for Darwinism is also an evangelist for atheism.[33] A conservative political strategist publicly embraces Darwinism because to him it implies the existence of natural limits to social progress.[34]

To limit the argument to the transformation of species and fail to engage the broader cultural issues of Darwinism is to miss the point. The transformation of species is just the tip of a Darwinian iceberg, and the iceberg is being generated by the scientists themselves, not by the masses.

The responsibility, consequently, lies with the scientific community to disentangle the issues. And they don't seem to be terribly interested in doing so, any more than they were in the 1920s. A poll conducted in 2005 found that the United States ranked thirty-third among sampled nations in the percentage of people who "believe in evolution," falling in between Cyprus and Turkey.[35] The conclusion was presented as evidence for the prevalence of "science illiteracy" in the United States, but it would hardly have surprised H. L. Mencken, who long ago dubbed the American public "the booboisie."

There is a different way of looking at the study, however. Simply from the standpoint of considering science, I find it troubling that scientists actually care what people "believe in." I would not be very enthusiastic about the legislative or judicial community taking such an active interest in what I believe, and I don't feel any better about the scientific community taking an interest in it. After all, the principal reason for doing so is to exercise ideological power. As the Inquisitors knew, in order to prosecute heretics, you have to identify them first.

What we need is not to condemn the masses for their ignorance or rejection of evolution (and presumably then, to make them believe in it). Rather, we need the scientific community to differentiate for the booboisie unambiguously between the rational, naturalistic study of life—which has led to the recognition of an ancient earth, a succession of living forms, and the genealogical connections among them—and the crap that invariably also gets attached to Darwin's name.

The task would be a lot easier if scientists didn't have their own agendas to push along with Darwin's. But that is exactly the point: Those more parochial "Darwinisms" are weakened by decoupling them from Darwinism, and consequently it is rarely in the scientist's interests to make those critical distinctions for the public. And in blaming the ignorance of the masses, rather than their own shortcomings, scientists can deflect the responsibility for the obvious gap between scientific knowledge and public beliefs.

Nevertheless, as citizens and as scholars, we are constantly making decisions about what science to accept and deal with (e.g., germs, global warming, reproductive technologies), what science to reject (e.g., racist physical anthropology, sexist evolutionary psychology, medical claims about cigarettes sponsored by tobacco companies), and what science to ignore (e.g., quantum electrodynamics, helminthology, nearly anything a biologist has to say about religion).[36] Nobody accepts with equal value all things presented with the authority of science, and one would have to be almost unbelievably foolish even to try; at all levels we exercise critical judgments, based on our experiences, education, and values.

Anthropologists have long known that religious systems serve many complex functions: intellectual, social, and politico-legal. Marriage is sanctified; God saves the queen, blesses America, and enjoins the Protestants to oppose the Catholics, the Shiites to oppose the Sunnis, and the orthodox Jews to oppose the reform Jews. The nominal segregation of religion from other facts of human life is the odd invention of a modern secular society. Religion, however, is where human beings generally find answers to questions about family, appropriate conduct, death, justice, and the meaning of life. Religion binds a person to a community of the like-minded; religion distinguishes right from wrong and enjoins you to embrace the former and to reject the latter.[37]

And what ties these diverse cultural aspects of religion together is the deployment of their origin myths. Origin myths not only tell you how you came to exist, they tell you *why* you came to exist, where you fit in, and why you should be good.

Now consider what bioscience has asked of the public for the past century and a half. It imagines that, far from being a component of an organic cultural system, an origin myth can be treated as a module that can be snapped out and replaced—like a Lego—without affecting other elements of the cultural system. That is exceedingly poor anthropological thinking. People who want to find meaning in their lives are not stupid, merely human.[38] And if evolutionary biology does not provide it, then it seems hardly fair to regard the people as deficient, when the deficiency is science's. This is not to say that science should try to render people's lives meaningful—a service for which there is pretty much a universal demand—but rather that science cannot reliably provide this service. When it has tried, to crib once again from H. L. Mencken, "it ceases to be science and becomes a mere nuisance, like theology."[39] Consequently, the endeavor tends to debase what science can actually do well.

Worse yet, science commonly arrives with the evangelical message that religion is dead, and that we should all become scientists now; if you resist any part of its own convoluted message, you are "anti-science."[40] Of course, no one is anti-science; that label smacks of paranoia. Actually, even the most sociopathic hackers and fire-and-brimstone televangelists accept a considerable amount of science simply in the pursuit of their interests. If anything, even the "scientific creationists" of the 1970s were pro-science; they were seeking legitimization from it and thus ironically privileging science as an authoritative source of knowledge.

So let us, once again, step back and take an anthropological look at the situation. We are busily delegitimizing the idea that our species was created from nothing by miracle in the Garden of Eden and saying instead that we arose naturally from monkeys.[41] That is more accurate, but less meaningful. Does it tell us about good and evil? Does it make us feel good about things we can't control? Does it affirm that we live in a universe governed by benevolence and justice?

Obviously not. Yet those are probably more important questions to most people than whether or not they came from monkeys. What might be interesting to know is this: given that nobody accepts or rejects all science, then how do people draw the boundaries between acceptable and unacceptable science? If someone else's beliefs are important to you, then you might try to see what it takes to get them to redraw those boundaries. The alternative—ridiculing people or bullying them into accepting your beliefs—rarely works. It's far more likely to make people resent you and resist you; any anthropologist knows that.

SIX Bogus Science

A distinguished biology professor at Yale writes in to a respected journal to defend a friend of his against the charge of scientific fraud: "The . . . work is good science inasmuch as it is repeatable and independently corroborated."[1]

Well, maybe the conclusions were repeatable and corroborated, or maybe they weren't, but is that really what constitutes good science? There are two nested problems here: the misplaced emphasis on the conclusions, rather than on the methods, as the crucial indicator of the quality of the science; and the casual manner in which a science educator can so grossly misrepresent the scientific endeavor to readers, so that it might actually reflect a normative view presented in science classes in the United States.

The specter lurking behind this queer statement is the prospect that generations of science students have been learning that getting the right answer is somehow more important that doing the work carefully and honestly. That is, after all, what the biologist said: corroboration is what makes for "good science," and consequently that corroboration will trump the accusation of data falsification. The problem is that the corroboration is *not* the work under scrutiny; it is the product of someone else's labor and, consequently, not a reliable guide to the quality or honesty of the original work.

Placing a premium on the result distorts the nature of the scientific enterprise, by placing a strong value on getting the right answer—an interest that may well conflict with honest reportage. This conflict could have fairly benign results, but it also is very readily manifested in the most basic ways.

HUMAN CHROMOSOMES

A very small fraction of the cells in the human body are actually dividing at any given time, and chromosomes are observable only during cell division. A hundred years ago, studying them involved making thin slices of human tissue, fixing and staining the cells, finding cells that had died while dividing, and counting the chromosomes. Since the chromosomes were entangled among themselves, the latter task was roughly equivalent to counting the strands on a plate of linguini.

Between 1880 and 1920, this task was attempted by a substantial number of cell biologists, whose results ran an extraordinary gamut. Before it was appreciated that at death chromosomes rapidly clump together, and consequently some fixation of the tissue is required, human chromosome counts were quite low, often around twenty-four.[2] The prominent exception was a Belgian cytologist named Hans von Winiwarter, who insisted on using carefully and freshly fixed testicular preparations for his analyses. Beginning in 1912, he began to argue that the established number of human chromosomes had been substantially underestimated. The high quality of his cytological preparations made his arguments difficult to dismiss. Males appeared to have forty-seven chromosomes. These could be harmonized with a known sex determination mechanism in some insects, wherein the male was XO and the female XX; thus, human females could be inferred to have forty-eight chromosomes, which is indeed what Winiwarter counted.

One possible resolution of this discrepancy relative to the earlier, lower counts was that Winiwarter was counting the diploid number (in body cells) and everyone else the haploid number (in reproductive cells). Alternatively, since several of the counts of twenty-four had come from scientists in the United States examining the cells of black men, it was possible that blacks had half as many chromosomes as whites. As the geneticist Michael F. Guyer wrote in *Science,* "I am at present engaged in a study of material from two different white men and although not yet ready to make a detailed statement I can say with assurance that the number of chromosomes is considerably in excess of those found in my negro material."[3]

In the early 1920s, however, a major figure emerged in American cytogenetics. Having studied the meiosis of spiders for his doctorate at Yale, Theophilus S. Painter established that opossums have an XX/XY system of sex determination and then set to work on humans. Studying quickly fixed cells from the testes of castrated black and white inmates of a mental institution in Austin, Texas, Painter found that whites and blacks did not differ substantially in chromosome number and that "the diploid number of chromosomes for man is very close to the number (47) given by Winiwarter." However, since Painter was satisfied from his meiotic studies that humans had an XX/XY sex determination system like the opossum, he would "expect an even number, that is, either 46 or 48."

Initially Painter's data suggested forty-six to be the right number. The key, however, was in the sex determination system. Painter reasoned that Winiwarter had been misled, not by poor technique, but by both his commitment to the XX/XO system, which implied an odd number in human males, and the fact that the Y, being small, is easy to miss. "Thus he added his sure Y to Winiwarter's sure 47 to get, in his opinion, a sure 48 chromosomes in human cells."[4]

By 1925, that number was firmly established in tables and textbooks. What is particularly interesting, however, is that once the "right" number was established, it became much easier to find. Immediately, Guyer (whose previous chromosome counts had been in the twenties) himself was counting forty-eight and finding the X and Y. And for the next thirty years, not only was this figure established as correct, it was continually replicated by subsequent investigators.[5]

As it turns out, however, humans do not have forty-eight chromosomes. The scientists were finding and counting their expectations. Utilizing three technological advances (growing cells in suspension, adding chemicals that prevent cells from finishing cell division, and squashing rather than sectioning cells), Tjio and Levan examined 261 cells from twenty-two cell cultures derived from the lung tissue of four human embryos and found forty-six chromosomes far more frequently than any other number. This surprising discovery in 1956 was supported immediately by gametic counts.[6]

Somewhat embarrassed in retrospect by his own consistent miscount of forty-eight human chromosomes a few years prior to the discovery that there were only forty-six, the cytogeneticist T.C. Hsu, who had developed some of the technical improvements, likened himself to a football player who returns an interception forty yards only to fumble at the three-yard line. Hsu recalled attempting to stretch his chromosome counts out to the number he "knew" was correct, forty-eight. And by now, T.S. Painter was not only an icon of cytogenetics but president of Hsu's university as well. "It was unthinkable that Painter could be wrong." Indeed, Tjio and Levan themselves wrote of a group of their colleagues who had abandoned cytogenetic studies of liver cells after consistently counting forty-six and thus failing to find the nonexistent last two chromosomes they had expected to see.[7]

REPLICATION

For decades, scientists were seeing the number of human chromosomes that they expected to see, because (1) they couldn't really distinguish forty-six from forty-eight with the quality of the data at their disposal; and therefore (2) they couldn't show forty-eight was wrong; so (3) it might as well be forty-eight, which was the official number. More generally, the problem was that they had valued precision over accuracy. What they could tell reliably was that humans had a number of chromosomes in the high forties, but somehow that did not sound quite scientific enough. It was better to produce an integer—even if they couldn't really prove that particular number was right—than to acknowledge that they weren't exactly sure what the number was, although they could get pretty close. So they ended up seeing, again and again, the wrong number.

This brief historical anecdote illustrates an important point about replication of scientific results: there are several reasons why different researchers might get the same answer, only one of which is that the answer itself is actually correct. That is why another researcher getting the same answer has no bearing on whether the first researcher has committed fraud. If this seems like belaboring an obvious point, then turn it around: the point apparently sometimes needs to be made to science professors at elite educational institutions.

Replication is easily overvalued in science. In the first place, very little replication actually goes on in science, for scientific funding is driven by the perception of novelty, and novelty is the very opposite of replication. Second, in cutting-edge research there is always enough technical skill required that a failure to replicate can be effectively countered by saying they just did it wrong.

What replication does give us is a coarse screen for artifactual results. At the most basic level, if the first scientists made an idiotic mistake, then having someone do it over might show how they got the weird answer they did. Of course, that presupposes that someone cares enough about the result to challenge it, which implies that the work itself is at least minimally controversial. And if it takes an expensive set of equipment or obscure knowledge, then replication might be, at best, extremely difficult.[8] But it can work. A failure to replicate an experiment or analysis can indeed lead to a correction.

In the early 1950s a bizarre exchange took place between two highly respected anatomists. In one corner was the Oxford anatomist Wilfrid E. Le Gros Clark, later knighted; and in the other corner was his former assistant, the Birmingham University anatomist Solly Zuckerman (who died in 1993 as a baron—that is to say, Lord Zuckerman). Le Gros Clark had become convinced that the South African australopithecine fossils were those of a human-like creature. Zuckerman maintained they were apelike, and for the coup de grâce he demonstrated with newfangled statistics that the teeth of australopithecines in particular were more like those of apes than like those of people. The problem was that it is obvious, from just looking at australopithecine teeth, that they are more like human teeth (with small canines and relatively large molars). Le Gros Clark got himself a statistician, and in 1951 they showed that

Zuckerman's calculations simply had to be wrong—and Zuckerman publicly acknowledged that he had neglected to divide by the square root of two at a crucial juncture.[9]

Examples like that—where someone tries to redo it and doesn't get the same result, which leads to the original authors realizing that they did something incorrectly and eating crow—are, however, very rare in science. More commonly, small differences in the protocols are magnified, heels are dug in, positions are reiterated, competencies impugned, and a "controversy" develops or continues. And, let's face it, couldn't the failed replication just as easily be the one in error?[10] In fact Le Gros Clark's statistician had used Le Gros Clark's measurements, not Zuckerman's, and had used a more powerful multivariate, rather than a univariate, statistical analysis. It is to Zuckerman's credit that he acknowledged the mistake.

Replicability is a property of science, but it is actually just a specific expression of the more general social aspect of scientific knowledge and practice. Because science is a social practice—relying on other people's work more than on individual revelation—replication is built into the system. One person's science affects other people's science; so sooner or later mistakes are uncovered, as steps are retraced when science is confronted with an apparently dead end.

But what does a failure to replicate mean? It means, most likely, that somebody is wrong—but not necessarily the earlier work. Replication, as noted, is never perfect, which always leaves wiggle room for negotiation. Margaret Mead's 1928 best seller, *Coming of Age in Samoa,* showed that teenage Samoan girls had less adolescent sexual angst than their American counterparts, which directly implied that the feature was neither inevitable nor "natural." Several decades later, a pugnacious Australian anthropologist named Derek Freeman claimed that Mead had been lied to (and that he hadn't been), and that his work in a different part of Samoa after World War II falsified her ethnography from the 1920s. Freeman's claim was quickly embraced by like-minded biological determinists. The judgment of the more recent anthropological community is that the differences of space and time obscure the relevance of Freeman's work to Mead's own; that Freeman considerably overvalued the significance of Mead's 1920s work and his own; and that Freeman's ethnography is at least as full of holes as Mead's.[11]

But just as the failure to replicate a finding doesn't guarantee that the finding is wrong, neither does successful replication guarantee that the finding is right (much less, as already noted, that the original finding was honest). In other words, two successive researchers may both get the wrong answer. Why?

One reason might be that the replication was biased by the original study. The answer is always easy to find when you know what answer you're looking for. This presumably was the situation with the human chromosomes. The scientists consistently continued counting past the forty-six that were actually there to find the forty-eight they "knew" to be there. At the very least, it shows the bias introduced by knowing the result in advance.

(And all the cytogeneticists were doing was counting.)

Another reason that two researchers might get the same incorrect result is that they are observing the same artifact. A well-known example of this occurred in the late 1960s, as Soviet chemists synthesized a new polymerized form of water, with different properties, but the same structure, as good old H_2O. Different groups had different degrees of success in replicating it, and some urgency was even suggested by Kurt Vonnegut's 1963 novel, *Cat's Cradle,* in which a new form of water, called ice-nine, is part of an apocalyptic scenario. It seems as though polymerized water was being produced in such minute quantities that impurities leaching out of the labware, even when thoroughly cleaned, were giving the water its apparent properties. The different groups that replicated the finding were actually replicating the artifact.

Yet another reason that the same result might come up twice is that both cases were dishonest. Here we can consider the most famous case of scientific fraud, Piltdown Man, found in England in 1912. Combining apelike teeth with a human-like cranium, the fossil immediately raised questions about whether the teeth and cranium actually belonged together. But their discoverer, Charles Dawson, brought scientists to the site to look for more fossils and, sure enough, a canine tooth was found, and later, some other skull fragments and teeth. Whatever scholars were still sitting on the fence now accepted the association between the ape-like jaw and human-like cranium as real rather than as a fluke.[12] They were, however, wrong. The association was neither real nor accidental; it was fake and deliberate. Twice.

Replication, then, is not a guarantee of anything, and certainly not of good science. Replicated work can run the gamut from good science to incompetent or dishonest science and can thus hardly be taken as a defining attribute of good science. After all, the fact that more than one person has claimed to have received anal probes from space aliens does not mean they all really have. Good science is simply science done competently and honestly, nothing more complicated than that.

The other definition of good science—science that gets the same answer someone else got, which is ipso facto the correct one and which justifies the means of obtaining that answer (that is to say, the definition of good science that introduced this chapter)— is worth a closer look. The science historian Horace Freeland Judson calls this "a seductive idea . . . [representing] one of the deepest divides among practicing scientists today."[13] The problem is that replicated baloney is still baloney—and the issue is how to distinguish it from filet mignon, not how much of it you can produce. The looming implication is that a large percentage of practicing scientists may not have a clue about what good science really is or how to identify it—and you can draw your own conclusions about the merits of such people educating the public about it.

AND ON THIS SIDE, SHINOLA

The ability to make crucial distinctions is fundamental to any human activity. Isaiah 8:15 uses a metaphor of distinguishing butter from honey; more recently, anatomists have been known to mutter contemptuously about people who apparently can't tell their ischium from their distal humerus.

The ability to tell good science from bad science is a crucial distinction, and there certainly seems to be widespread confusion about it—even within the scientific community. One manifestation of the problem is the term *pseudoscience,* a word often thrown around as if it were easy to tell from the real thing. Once again, however, history shows that it isn't. Obviously anything promoted as science by nonscientists—that is, by people dubiously credentialed—is likely to be pseudoscience. But, actually, a lot of intellectual fads that are identified in retrospect as pseudoscience were promoted by scientists themselves.

In the mid-1800s, medicine had become professionalized, and scientific theories of human behavior were being sought (and still are, of course). Cranial anatomy seemed like a good place to look for such a theory. The skull, after all, encases the brain, which is the seat of thought, which is the basis for behavior. A significant bulge in the skull might well reflect a hypertrophied part of the brain, which in turn might imply an overdeveloped sense of whatever that particular part of the brain actually does for its bearer. Amorousness, inquisitiveness, trustworthiness, and musical sense are all mental states, are all ultimately products of the brain, and are all more or less developed in different people. Does it not stand to reason that a close study of the skull would allow us to identify those aptitudes in different people? Led by physicians and anatomists, the principal advocates of phrenology—J. G. Spurzheim and Franz Joseph Gall in Vienna, Samuel George Morton in the United States, Paul Broca in France, and George Combe in England—all claimed to be speaking for science, and who is to say they weren't? The scientific community was divided over the merits of phrenology, but that illustrates the point perfectly: why couldn't a significant chunk of the scientific community tell that it was nonsense? Half a century later the work could be dismissed,[14] but there has got to be a better way to tell pseudoscience from good science than simply waiting a few decades to see what people of the future think.

(And yes, people of the future are still stupid. Scientists in November 2007 correlated the feelings of potential U.S. voters with the glow of their brain parts and published the results in the *New York Times;* seventeen cognitive neuroscientists responded indignantly but acknowledged that there is a lack of quality control in this research area.)[15]

Nor is that example isolated. At the same time that the scientific community was finally rejecting phrenology as pseudoscience, they were embracing eugenics. Madison Grant's *Passing of the Great Race,* essentially a pre-Nazi primer of Nazi science, was reviewed favorably in *Science* by a geneticist from MIT, Frederick Adams Woods. The same reviewer would later publish a metareview of Grant's book in a genetics journal and argue that most of the criticisms were coming from Jews and other people whose opinions were conflicted because they were not members of the Nordic race themselves. Within two decades, however, the label *pseudoscience* would be casually applied to Grant's work.[16]

That presents this interesting question: how did Grant's work go from science—and apparently *good* science[17]—to pseudoscience? It's not as if geneticists and paleontologists finally found conclusive evidence that the Nordic race wasn't really so great after all. Obviously, rather, certain voices that were not previously being heeded were somehow finally being heeded. Did it have something to do with the passing of Grant's principal sponsor, Henry Fairfield Osborn, president of the American Museum of Natural History, in 1935? Did it have something to do with the wholesale embrace of Grant's work by the Nazis?[18]

Grant was not fully credentialed (he was trained as a lawyer), but he was acknowledged as a lover of science—an amateur in the literal, and favorable, sense. His obituary in the *New York Times* identified him as a zoologist, dwelt upon his work in conservation, and called *The Passing of the Great Race* "a recognized book on anthropology."[19] Serving under him, on the advisory board of the American Eugenics Society in 1927, was an impressive swath of biologists and geneticists: William E. Castle of Harvard, Wesley R. Coe of Yale, Leon J. Cole of Wisconsin, Edwin G. Conklin of Princeton, C.H. Danforth of Stanford, Edward M. East of Harvard, Michael F. Guyer of Wisconsin, Samuel J. Holmes of Berkeley, Ann Haven Morgan of Mount Holyoke, Horatio Hackett Newman of Chicago, Herbert E. Walter of Brown, Frederick Adams Woods of MIT (small world!), and Sewall Wright of Chicago. And, for good measure, alongside Grant on the board of directors was C.C. Little of Michigan, who would later found the mouse genetics center at Bar Harbor, Maine. Being appalled by Madison Grant, and placing as much distance as possible between his proto-Nazi ideas and your scientific ones, was a decidedly minority view among American biologists.

It is hard to avoid the conclusion, then, that the passage of Madison

Grant's work from good science to pseudoscience was the result not of discovery but of social process. No previously unknown facts of nature were unearthed, but previously known data and arguments began to seem more persuasive in light of social and political developments. To call *The Passing of the Great Race* pseudoscience today is to obscure the fact that most geneticists and biologists simply did not identify it as such over the first two decades in which it was in print.

Why not? How can you not be able to identify pseudoscience in your area? Wouldn't that have to be a minimal requirement for some sort of scholarly competence?

BAD SCIENCE

Oddly enough, the identification of bad, bogus, incompetent, or pseudoscience is generally not part of the curriculum of science students. How do scientists acquire such knowledge, then? Sir John Maddox, editor of the world's most prestigious science journal, addressed that issue in 1988 when he initiated an episode to show

> how easily authentic science may be simulated by the careful selection of data and the judicious use of language, how even "rigorously and fairly" reviewed papers may embody defects recognizable even by . . . "amateurs," and—more alarming—how likely it is that much second-rate science finds its way into print somewhere.

The implication is that science produces unreliable work with at least comparable prolificacy to its reliable work, so that whatever standards are in place to differentiate between them either don't work well or are not well enforced. How does one assess reliability, then? What Maddox did was to put the issue on the table by publishing an article by respected researchers (led by the French immunologist Jacques Benveniste, who sincerely believed what his group had done) along with an "Editorial Reservation," to wit:

> Readers of this article may share the incredulity of the many referees who have commented on several versions of it during the past several

> months. The essence of the result is that an aqueous solution of an antibody retains its ability to evoke a biological response even when diluted to such an extent that there is a negligible chance of there being a single molecule in any sample. There is no physical basis for such an activity. With the kind collaboration of Professor Benveniste, Nature has therefore arranged for independent investigators to observe repetitions of the experiments. A report of this investigation will appear shortly.

Not only was Maddox saying that he didn't really believe something that he was nevertheless publishing, but he neglected to specify the composition of the visiting committee. There would be no immunologists on the committee; it would comprise a physicist, a fraud buster, and a stage magician. The physicist was Maddox himself, who wondered how scientists can convince themselves that impossible results are not only possible but real. (In the first place, it helps not to think they are impossible; Benveniste's work had implications for the "alternative medicine" of homeopathy, in which infinitesimal dilutions of a potent substance are believed to have healing powers.) The fraud buster was Walter Stewart of NIH, whose principal interest was, very controversially, the integrity of the scientific literature. And the magician was James Randi, whose expertise lay in having mastered techniques of convincing people that they were seeing things they were not actually seeing—that is to say, magic.

Clearly the composition of the committee was intended to determine *not* whether wool was being pulled over the scientific community's eyes but whether it was alpaca or merino. Their report identified (1) a financial conflict of interest in which two of the authors of the article were employed by a company that manufactured homeopathic pharmaceuticals; (2) a poor set of laboratory standards that blurred the distinction between expected and anomalous results; and (3) the failure of the laboratory to produce reliable results when the investigator did not know what the sample contained beforehand. It ended with a parable on the difference of documentation required to establish two different claims: "I have a goat in my backyard" and "I have a unicorn in my backyard."[20]

Granted, not all labs are equal. But they are all nevertheless science.

The Benveniste work was highly technical and executed by qualified, expert scientists. How, then, do we tell science's more reputable products from its lesser? Do we place too much faith in the peer review system? If so, what do we replace it with?

Everyone has their peer review horror stories. The criticisms that arise are generally reviewer incompetence, reviewer conflict of interest, and the lack of incentive to the reviewer to be thorough and conscientious in what is essentially a pro bono consultation.[21] You'd be paying for an expert's opinion if the time belonged to your lawyer or your plumber. Why should you expect scientists to provide the service for free, when their time is as valuable as yours? Don't you think they have better things to do with their time?

Peer review is a gatekeeper, with its own limitations. Beyond peer review, however, we can acknowledge a few guidelines for identifying poor, bogus, or pseudoscience. Obviously none of these signatures guarantees the quality of the work, but as they arise they perhaps ought to arouse suspicions. The list that follows is modified from one first suggested by the physicist Robert Park.[22]

1. *Originating outside the ordinary channels of knowledge and expertise*

The maturation of a scientific discipline is actually its professionalization, the development of a canon of knowledge and a means of accrediting those who have mastered it. Those without appropriate credentials are not taken seriously, for they have not demonstrated expertise. If amateurs could regularly make useful contributions, there would be no difference between amateurs and professionals; astronomy would still be indistinct from astrology, chemistry would still be indistinct from alchemy, and you wouldn't be able to tell doctors from barbers and midwives. It would be an intractably relativistic situation.

But one of the reasons that science works is that some people have taken the time and trouble to become experts. Scientific knowledge is not just common sense or impressions; it is specialized, accurate knowledge—and some people just have more of it than others. This is what makes the contributions of nonspecialists generally suspicious: someone who doesn't really know what they're talking about—or reads everything with the utter credulity of the untrained mind—is simply unlikely to advance the field. The attention such work receives from the experts is likely to be dismissive, and rightly so.

In 1844, a Scottish publisher named Robert Chambers decided to put his ideas about the transformation of species before the British public. They were not received favorably, as Chambers had expected, which is why he published his book, *Vestiges of the Natural History of Creation*, anonymously. But the book was spectacularly successful. In 1854, Thomas Huxley would write of it, "Time was, that when a book had been shown to be a mass of pretentious nonsense, it, too, quietly sunk into its proper limbo. But these days appear, unhappily, to have gone by, and the same utter ignorance of the public mind as to the methods of science and the criterion of truth, . . . have encouraged the author of the 'Vestiges' to venture upon a *tenth* edition." And just for good measure, "We grudge no man either the glory or the profit to be obtained from charlatanerie."[23] Just a few years later, the same reviewer would champion the transformation of species, as elucidated more expertly by Charles Darwin.

Likewise, when the dramatist Robert Ardrey published *The Territorial Imperative: A Personal Inquiry into the Animal Origins of Property and Nations* in 1966, the anthropological community was unwelcoming. Sir Edmund Leach, the leading anthropologist in England, began his review, "As a mine of scientific-sounding misinformation Mr. Robert Ardrey would be hard to beat."[24]

Admittedly, it is hard to review such a work impartially, because its very existence implicitly demeans your own expertise. The premise of the work is that the experts are all wrong and a nonexpert is needed for correction—which is the basic anti-intellectual premise on which creationism is founded as well. Amateur theorizing is, a priori, simply unlikely to be of much value.

Counterexamples do, however, exist. In 1980, the Nobel laureate physicist Luis Alvarez proposed that paleontologists were wrong about the extinction of the dinosaurs and was greeted coolly by much of the paleontological community. Alvarez and his son, the geologist Walter Alvarez, had found that the strata around the time of the extinction of the dinosaurs were enriched with the element iridium, which tends to be present in higher concentrations in extraterrestrial objects, and thus suggested that the dinosaurs were wiped out by an asteroid or cometary impact. Being "plugged in" to respected academic channels helped give their heterodox ideas a hearing and ultimately led to a general acceptance of the possibility of a global ecological catastrophe triggered by an impact and a greater appreciation for the role of mass extinctions in the history of life.

2. *Ideological or financial conflicts of interest*

If that Pontiac dealer we met in chapter 2 were to tell you that a Pontiac is the best car on the road today, you would be unlikely to accept the statement at face value. You might hope that such a statement could be accepted as the uncompromised truth, but you recognize that the Pontiac dealer has a good reason for compromising it. That is not scientific expertise; it is simple common sense. People tend to say things that suit their purposes; that is why apes started talking in the first place, way back in the Pleistocene.

It is no different in science. A scientist with an economic interest in a body of work can hardly be expected to be impartial. Consequently, if somewhat belatedly, science journals have begun to insist on financial disclosure for articles they publish (not that there is much hope of enforcing such a rule). When the investigating committee visited the French laboratory that had claimed to validate homeopathy, they expressed their disappointment at the financial support given to the lab by companies producing homeopathic medicines. And like the police chief in *Casablanca* who was "shocked, shocked!" to see gambling in Rick's Café Americain before collecting his winnings, Jacques Benveniste replied indignantly, "Does the fact that homeopathic companies are paying two researchers . . . mean that they order them into improper conduct? We could not self-finance a long-term international cooperation nor the expenses of this large group of investigators. Did the source of the money influence their judgement? What a level of argument!"[25]

The argument, however, was not a scientific one; it was a commonsensical one. Moreover, as we noted in chapter 4, scientists have other conflicts of interest than simple economic conflicts. The stakes can be political, religious, or personal; being cultural animals, scientists can "invest" in a body of work several different ways. And obviously, the heavier the investment of any sort, the more unlikely it is that a scientist will be able (or desire) to evaluate it critically or impartially. Perhaps the most famous example of the use of political capital in science was the dismantling of Soviet genetics in the 1930s by Trofim Denisovich Lysenko. Invoking a version of Marxism that denied any limits on human social improvement, Lysenko concluded that "genes" necessarily constituted such limits and consequently must be a fabrication of bourgeois anglophone scientists. He began to develop a system of genetics without genes and, with the assistance of Stalin, had the most respectable Soviet geneticists jailed. The lesser Soviet geneticists kissed ass and did what they were told. Suffice it to say, Lysenkoism did not make any breakthroughs—at least in genetics. For students of the relations between science and society, however, Lysenkoism was a gold mine.[26]

The ideologies shaping scientific research may be subtler, as well. If you "know" that men are innately more aggressive than women, and maleness is governed by a Y chromosome, and you find a higher proportion of XYY men in mental and penal institutions than "out there," does it not stand to reason that that the extra Y chromosome is causing them to be extra aggressive, as geneticists argued in 1959? Actually it doesn't; the cause is the reduced intelligence that accompanies all forms of human aneuploidy (deviations from forty-six chromosomes), geneticists argue today.[27]

Where scientists enter into areas further afield of their background and training and have only technical expertise to guide them, they naturally tend to be unaware of the issues peculiar to their new field and of the intellectual nuances missed by an untrained mind. Bruce Lahn, a respected molecular geneticist at the University of Chicago, decided to delve into the mysteries of human evolution in 2004. Since brain size has been a significant factor in human evolution, he chose to study two genes associated with microcephaly, a pathological condition of premature cessation of brain growth. (Problem no. 1: Pathological variation does not translate well into an understanding of normal function. A broken oven precludes making a good beef Wellington, but a recipe for beef Wellington makes no reference to the condition of the oven. In other words, there are many things that can prevent a system from functioning properly that may be irrelevant to understanding the system itself.) From their pattern of differences when compared to the other primate homologs, Lahn inferred that the brain-growth genes had been under strong positive selection, presumably for intelligence. (Problem no. 2: The functions of the genes are unknown, and they are expressed in nonneural tissues.)

Then, for some reason, Lahn decided to survey modern human diversity for the two genes—and, sure enough, he found that the pattern was again consistent with the spread of a specific favorable allele, one coarsely dated to thirty-seven thousand years ago and the other to fifty-eight hundred years ago. And different populations have different proportions of them, with Eurasians having a higher proportion of one allele and Africans another. (Problem no. 3: Sixty populations, averaging twenty people per population, summarized the entire global distribution of the alleles.)

Lahn concluded that the genes were producing "different phenotypic outcomes of the brain." (Problem no. 4: Simple genetic variation does not readily translate into complex phenotypic products, like thoughts and deeds.) And maybe one of the genes was even related to the origin of food production and writing in the Near East. (Problem no. 5: Trying to explain social history in terms of genetics has been debunked as consistently as young-earth creationism.)

So, since modern civilization is ultimately founded on food production and writing, might that explain the cultural backwardness of Africans? And, perhaps as well, their position on the IQ "bell curve"?

Lahn never exactly said that last thing in the scientific literature. Others heard that message, though. But let's back up a second. Why was he studying human diversity in brain genes? What did he think he was going to find? Did he not know that explaining the products of social history by recourse to differences of nature is a political stratagem?

If he didn't know that, he was frankly too dumb to be working on problems of human variation. And if he did know it, then how could he be surprised when his political assumptions were called into question? After all, what really was the point of trying to relate these brain genes to "two important events in the cultural evolution of Eurasia"—even if "the significance of this correlation is not yet clear"?

Lahn was actually laboring under the popular racist axiom that Africans suffer from some intellectual deficiency, and, further, that a reasonable explanation for it is that they have inferior brains, which might be caused by a genetic peculiarity. For Eurasians, "Dr. Lahn favors the idea that the advantage conferred by the mutations was a bigger and smarter brain," said the *Wall Street Journal.*

But what, then, does it possibly mean that one of the brain genes turns out to have the same gene pool profile in the indigenous inhabitants of Papua New Guinea, Sardinia, and Mexico? Or that the other yields the same profile in the indigenous inhabitants of Brazil, southern Africa, and Cambodia?

Subsequent research, of course, failed to find any link between the mutations and either brain size or IQ—much less, cultural history. And Lahn's basic narrative explanation for his data—selection driving the spread of good brain genes—turns out to be no more likely than the patterns being a consequence of the basic ebb and flow of human populations, that is to say, the vagaries of demographic history.[28]

The point is that, whenever anyone tries to explain the social facts of history in terms of genetic microevolution, it's a good bet that they don't know what they're talking about—even if they do know a lot about sequencing DNA (or in the case of the Stanford physicist William Shockley in the early 1970s, about transistors). The damage done, Lahn retreated from this line of research, shocked—shocked!—at how politicized *others* had made it. The simple fact is that this science is indeed fundamentally politicized, and the scientist who ventures into it thinking otherwise—that is to say, knowing virtually nothing about the field and being unable to think critically about his own assumptions—is insufficiently self-aware to be taken seriously. He is essentially a talking horse, and a racist talking horse at that.

Perhaps, with different brain genes, Lahn would have produced a higher caliber of science—but somehow I doubt it.

3. *Bypassing ordinary conventions of legitimization*

As the investigating committee tried to explain to Dr. Benveniste, extraordinary claims have to answer to extraordinary standards. Those standards are imposed by the experts, the gatekeepers. One of the radical breaks taken by some early scientists—notably, Galileo and Descartes—was to write in the vernacular rather than in traditional academic Latin. This made their work accessible to a wider readership and, in effect, bypassed the academic gatekeepers.

Today the gatekeepers themselves are scientists, but they still have a gate to keep. And they keep it through the exercise of their judicious faculties for determining what enters the scholarly literature and what doesn't: peer review.

But suppose you have a really cool idea and don't want to subject it to the possible anonymous humiliation (and scientists *can* be very humiliating, especially when their name is not attached). Well, you can follow the example of Galileo and Descartes and take it to the masses.

Of course, the gatekeepers won't like that. They'll respond by vigorously denying the legitimacy of your claims—and all other things being equal, they'll have a point. If you could have published it unproblematically through ordinary channels, you would have. The reason you bypassed those channels was obviously because you felt you needed to.

A famous example of this came in a potential source of cheap energy, cold fusion, announced at a press conference in 1989 by two chemists from the University of Utah, Martin Fleischmann and Stanley Pons. It didn't really exist, but if it did, it would be a bonanza for the authors and their institution. So they staked their claim without allowing the scientific community to judge the merits of their research beforehand. Perhaps if their claim had actually been true, they would have been judged less harshly—but ultimately the claim was disproved (after some initial false claims of replication) and the chemists judged harshly for talking to the press before demonstrating the validity of their claims.[29]

Some of the leading journals have instituted embargoes on articles in press, issuing releases so that journalists can be prepared when the article is actually published but forbidding authors to talk to reporters until then. Again, however, enforcement is difficult. When Dolly, a sheep, was cloned in 1997, *Nature* sent out an embargoed press release, saying, "You can't write this up until Friday," but after the *Observer* in England wrote it up prematurely, the *New York Times* did as well, and *c'est la vie.*[30] In that case, however, Ian Wilmut, the principal scientist, was playing by the rules and not actually trying to circumvent ordinary channels. The news was just too big to keep bottled up, and Wilmut ended up as *Time* magazine's "Man of the Year."

4. *Overinterpreting a small scientific effect at the very limits of detection*

Pretending that data are better than they really are is one of the best giveaways, but it is easy to see only in retrospect. It was the most basic problem with the false human chromosome count and as well with the pseudodiscovery of polywater. Wishful thinking can make the difference between an unconvincing, unpublishable result and a barely convincing, publishable one.

"Report suggests homosexuality is linked to genes" ran a heading on the front page of the *New York Times* on July 16, 1993, and indeed it was true. Such a report had just been published in the journal *Science* (not *Pseudoscience*), led by the NIH molecular geneticist Dean Hamer. Like the XYY work earlier and Bruce Lahn's work a few years later, Hamer's work combined the highest-tech molecular data collection techniques with some very primitive assumptions about general genetics and human behavior.

The paper showed that in forty sets of gay brothers, genetic markers on the tip of the X chromosome (Xq28) seemed to match each other more frequently than expected by chance. Chance was 50 percent (since brothers inherit one of their mother's two X chromosomes), but this group of gay brothers matched X chromosomes in over 80 percent of cases. On the heels of a (now rejected) claim about the distinctive brain anatomy of gay men, Hamer's work seemed to have identified a real gay DNA segment.

But it hadn't. In the first place, there was not a particular common DNA segment that the brothers in Hamer's 80 percent shared; rather, each man simply had a segment of X chromosome that matched his brother's more frequently than would have been expected. In the second place, the field of gender and sexuality studies had little use for essentialized or innatist explanations of human sexual practice, for it was busy documenting the diversity and ambiguities of its subject across societies and eras. Third, geneticists generally are skeptical of the easy mapping of human genes onto nouns—which represents an ancient fallacy dressed up in Mendelian terms. A gene "for" sexuality is like a gene "for" an elbow: the trait is there and develops as a result of the expression of genes, but the genetic units and the words don't correspond.

But, most important, the very nature of the research (looking for correlations between genetic markers and behaviors) produces lots of "false positives"—statistical associations that are actually spurious. John Maddox, editor of *Nature* (Hamer had published in its rival, *Science*), observed that his own journal had only recently published similar associations about schizophrenia that quickly turned out to be spurious. In fact, this line of research is very sensitive to the nonrandom elimination of data. It takes only a few discordant families to be tossed out—for whatever reason—for the association to attain statistical significance, which means "publishable."

Hamer's group even claimed to replicate the findings a few years later. But after an independent study failed to find the same association and the Human Genome Project failed to reveal anything of great interest in the DNA sequence of Xq28, the "gay gene" quietly sank into oblivion.[31] Hamer continued to identify imaginary genes for properties like "novelty seeking" and "belief in God" and to write popular books about them, thus disproving the widely held notion that the public needs to read more books about science!

5. *Relying on anecdotal, experiential, or hermeneutic data and methods*

The social nature of science is its most distinctive feature, and replicability is one facet of that social nature. If a claim requires that "you had to be there," then it is not a scientific one—although it may well be about something far more interesting. Science inevitably requires experiencing and interpreting, but there has to be more to it than that. Experiencing and interpreting are internal and idiosyncratic acts, and thus are themselves of limited value to science, unless the community can somehow share in the experiences and be convinced that the interpretations are the best ones possible.

While Freud may have been mistaken about some things, he was certainly right that, for example, dreams reveal unconscious anxieties. They are neither portents of the future (like the pharaoh's in Genesis 41) nor instructions from Beyond (like Joseph's in Matthew 2) but the human mind's attempt to express and resolve conflicts in a symbolic form. The problem lies in knowing which interpretation is right.

Interpretation is necessarily the most crucial act of any scientific enterprise, for data are valuable only insofar as they are meaningful. But since multiple interpretations of data are always possible, there are few guidelines available for evaluating the proper one. We can say that, on the whole, scientific explanations should tend to be mechanistic, as opposed to teleological or purposeful; naturalistic, as opposed to spiritual; parsimonious, as opposed to cumbersome; probabilistic, as opposed to deterministic; and logical, as opposed to inconsistent or irrational.

And yet exceptions are rife. Isaac Newton's revolutionary theory of gravity consisted of describing an invisible force binding two objects together, hardly a modern-sounding scientific proposition. Rudolf Virchow's revolutionary cell theory, that all cells arise from other cells, is more compatible with the idea of vitalism (that life itself has a special nonmaterial quality) than with the materialistic alternative of spontaneous generation (that life is continually arising, since it is continuous with nonlife). Yet vitalism seems to be wrong, in spite of the cell theory; and so does spontaneous generation, in spite of its materialist underpinnings.

Likewise, any scientific study of human behavior has to begin with the recognition that, although there are some continuities between what humans do and what monkeys do, human activity is purposeful in ways that monkey behavior is not and thus requires a more complex theory than simply a variant theory of monkey behavior.[32] At its most basic, the question "What were they trying to do?" can be addressed for human behavior (by asking them), while for monkey behavior it simply can't be.

So sometimes you need invisible forces, discontinuity, and complexity—much as you want to avoid them in scientific explanations—because sometimes they are right.

This discussion presupposes that the interpretive, hermeneutic aspect of science is accompanying a fairly straightforward interpretation of data, within the general social framework of scientific expertise. But suppose, instead, that the hermeneutics is all there is. If it comes from a community member, it can be casually dismissed as "hand waving." But if it comes from outside the community, then it effectively constitutes an anti-intellectual challenge to the community's scientific authority, and it becomes *uninformed* hand waving (see no. 1, above). Worse yet, if it gets into the public eye (see no. 3, above), the task of public science education is set back, and the masses may start believing things they shouldn't.

And we can't have that, can we?

Immanuel Velikovsky was a medically trained psychoanalyst with peripatetic interests. One such interest lay in the Enlightenment project of naturalizing miracles. If we maintain simultaneously that the Bible is the word of God, and that God made the world to run on laws, not miracles—then how do we account for the miracles reported in the Bible? The most reasonable answer is that they were not miracles per se but were misunderstood as such by the bucolic yokels who wrote the events down. And if you believe that, then the next question is "What were the miracles *really*?"

In fact, there is a ridiculously copious literature that tries to explain

Biblical miracles as misunderstood natural phenomena. One example is the "Star of Bethlehem," which (according to St. Matthew but none of the other gospels) led some wise men from the east to King Herod's door. From there, "it went before them, till it came and stood over" baby Jesus (Matthew 2:9).

Gosh, what kind of a celestial object could do that?

Probably only a flying saucer. But if we focus just on the less idiotic parts of the story, we can pose the question "What kind of interesting astronomical phenomena might have been interpreted as a portent for the birth of the King of the Jews, recorded in Matthew?" Well now, that depends upon when you think Jesus was born. We know Herod died in 4 B.C., so it would have to be earlier than that (ignoring the fact that Luke's gospel has Jesus born in A.D. 6, during the census of Quirinius). In the early seventeenth century, the astronomer Johannes Kepler calculated that a rare conjunction of two planets occurred in 7 B.C. and that this might have been the Star of Bethlehem. Given the quantity of phenomena available, and the decade or so in which to place them, it turns out that nearly every generation of pious astronomers since Kepler has come up with some kind of natural phenomenon—conjunctions, supernovas, comets, meteor showers—to "explain" the Star of Bethlehem.

But, of course, the only evidence we have for the Star of Bethlehem is St. Matthew's account (so the Bible itself goes three gospels to one against it!), and according to classical sources the births of pretty nearly all the great figures of the ancient world were heralded by celestial events—that is to say, by symbolic astrological, rather than astronomical, phenomena. So there probably isn't even a Star of Bethlehem to explain in the first place. The modern astronomy itself may be sensible, but the astronomers' confusion of myth and history is not, and that is ultimately what renders the whole enterprise sterile.

It is within this interpretive tradition, however, that Velikovsky wrote *Worlds in Collision,* in 1950.[33] Let's assume, argues Velikovsky, that the events recorded in Exodus or the movie *The Ten Commandments* are sort of true. Obviously they didn't happen exactly as we have them written down, because we don't believe in miracles anymore. The Nile didn't really turn to blood, the eldest Egyptian child in every household wasn't precisely singled out for death, and the Red Sea didn't really part just to let the Children of Israel through. But something must have happened to be interpreted in that fashion, right? So, what natural phenomena could the author of Exodus have been recording somewhat imperfectly, so many years ago?

Velikovsky devised a clever theory to account for Exodus in which astronomers and physicists are simply all wrong about the history of the solar system.[34] In this scheme, only a few thousand years ago the planet Venus came into existence from the Great Red Spot of Jupiter. It knocked Mars out of its orbit, sending it very close to earth—affecting particularly the ecology of the Nile at the time that Moses was engaged in negotiating the Israelites' freedom with the pharaoh. The iron oxide on the surface of Mars fell heavily upon the Nile, making it undrinkable and resembling blood—if your blood looks like rusty water. The frogs (plague no. 2) didn't like it either, so they hopped out onto land, where they died, which precipitated the plagues of lice and flies (plague nos. 3 and 4). The contact of the planets led, directly or indirectly, to the rest of the plagues as well, and later to manna from heaven. And for good measure, before settling into its present orbit, Venus made another pass a generation later and stopped the earth's rotation for twelve hours as Joshua was laying siege to Gibeon. Finally, argued Velikovsky, this scenario could be verified by close readings of mythologies of diverse peoples. Could not the birth of the goddess Athena, sprung full blown from the head of Zeus, be an allegorical recollection of the planet Venus emerging from the planet Jupiter?

Astronomers were outraged and responded by threatening a boycott of the publisher's textbook division. Macmillan transferred the book to a company without a textbook division—but the selling power of a book that the scientific community had tried to suppress was more than worth the inconvenience. Of course, it did not pass unnoticed that Velikovsky's ideas had not been vetted by the ordinary process of scientific peer review (see no. 3, above), but trying to suppress its publication seems not to reflect the most admirable qualities that science tries to project about itself.

After its initial release in 1950, *Worlds in Collision* slowly faded away, until being rediscovered by the nascent "new agers" in the 1970s. *Worlds in Collision* was indeed "Occult" or "New Age" before those sections even existed in bookstores. The second round of interest in Velikovsky stimulated the American Association for the Advancement of Science to sponsor a symposium bashing him. Although Velikovsky spoke in the symposium, its published proceedings failed to include his defense, thus contributing to the image of the visionary genius being persecuted by the powerful, intellectually corrupt, conservative institution. Velikovsky showed no reluctance to compare himself with Galileo and Giordano Bruno, hounded for their heliocentrist heresies back in the seventeenth century.[35]

The oddest part of all this was that the scientific community consistently chose as the battleground the authority of their astronomical pronouncements—that is to say, the question "Whose history of the solar system is right?" They were right and Velikovsky was wrong, period.

But that handling of the situation not only legitimized Velikovsky's views (by making them worthy of attention and refutation); it also reinforced the image of him as an intellectual martyr. The arcane knowledge it took to refute Velikovskian astronomy was not relevant for much of the public, who just found it to be a cool idea. Velikovsky was challenging the power structure, a little guy being squashed by "the establishment." The defensive, if not paranoid, reaction of the scientific community was eloquent in this context. It produced exactly the opposite reaction intended, actually raising the questions "What's wrong with listening to him? What is the scientific establishment so afraid of?"

The problem is that, by addressing the wrongness of Velikovskian astronomy, science legitimized it as science rather than contrasting it to science. What was needed was to delegitimize not so much its false claims, but its source of inane knowledge production. That source lay in the assumptions that myths are imperfect recollections that can be recovered and that the interpretative recovery process can take precedence over scientific knowledge when they conflict.

The first assumption may hold, but only rarely. Myths are stories that people tell because they resonate with some aspects of their identities and feelings. Three millennia from now, when mythologists study the relationships of Anakin Skywalker to his twin offspring, Luke and Leia, they will be smart enough to know that Anakin is a character in a drama about good and evil, love and hate, power and loyalty, mechanization and intuition, and the search for self-realization in a large, alien, and unpredictably hostile environment. There is no reason to think Exodus is anything other than a story with grandly resonant themes, to explain to the Jews who they are and why they are special—in the context of remote circumstances removed from their actual lives by many centuries. Whether there really was a chap named Moses who received ten engraved commandments on the way out of Egypt is less important than the feelings that the story evokes. Indeed, to focus on its possible empirical content is to miss the point of the myth entirely and to misunderstand the nature of myths altogether. Mythmakers create their stories out of the available motifs and adapt them to the needs and interests of their community.[36] There is consequently no reason to think Moses was any more or less real than Odysseus or James Bond (both of whom, in a sense, delivered their people from the clutches of evil potentates, like Priam and Dr. No, and did so in the context of real events—the Trojan wars and the cold war). That doesn't make Moses any less important, or the Jews less coherent, or the Torah less relevant; it simply keeps the focus where it ought to be, on the story and its meaning.

Velikovsky's second assumption is even more problematic. Not only is he fabricating history from his idiosyncratic analysis of myth, with no clear methodology for doing so, but he is then privileging his own story over all other sources of knowledge, from the meanings of other myths to celestial mechanics, which must all now be interpreted to conform to his story.

When seen this way, Velikovsky's story about the cosmic billiard balls hardly requires refutation. It is anti-intellectual and egotistical nonsense. His is simply not a manner of producing scientifically reliable knowledge. Exodus is not a historical document any more than Genesis is; if there is any reality underlying the story of Exodus as we have it, there is no reason to think Velikovsky has found it. The myths of other peoples are not dim memories either and consequently should not be understood as such, much less twisted to fit this scenario. Nor is there any reason to twist the cosmos to conform to this idea. Consequently, the story in *Worlds in Collision* lies outside history and science, and outside astronomy in particular. It is myth analysis, and incompetent myth analysis at that.[37]

6. *Complaining of a conspiracy of suppression*

Wackos are in a double bind. They clearly see a conspiracy to suppress their ideas: science is an authoritative voice and, like other authoritative voices, is not terribly interested in hearing challenges to that authority. But people who call attention to that fact immediately situate themselves as wackos, that is to say, on the outside looking in on science. Of course there is a conspiracy afoot to suppress their ideas—if they are not credentialed, then chances are that their ideas are uninformed and of little potential value. Robert Chambers knew it, Immanuel Velikovsky knew it, and Robert Ardrey knew it. They shepherded their attempts to reform scientific thought into print, got some support from the masses, and took a beating from the scientific community.

Not too long ago, I found myself speaking to a group of AIDS researchers (who knew I am an anthropologist with principal interests in human evolution that don't overlap theirs but nevertheless wanted me to talk to them). Before me, another plenary address—this one by a very distinguished AIDS researcher—cited in passing a dilettantish book on human evolution from a few decades ago that is taken seriously by nobody in the field.[38] A bit later, it was my turn to speak. On the one hand, I didn't want to embarrass the previous speaker, but on the other hand, my job is to teach people about anthropology, and I can't very well let this group think that the work the other speaker had cited is actually highly regarded by the cognoscenti. What I said was simply that, for an anthropologist to hear that work being cited would be like me speaking to an AIDS conference and quoting Peter Duesberg. That was enough; they all knew what I meant.

Peter Duesberg is a distinguished molecular virologist from Berkeley, elected to the National Academy of Sciences in 1986. In the late 1980s, after years of neglect, AIDS was finally coming onto the scientific radar, and the cause had lately been discovered: a retrovirus called LAV by a French group and HTLV-3 by an American group (a decision was brokered for them to share credit and call it HIV). Much still was not known about it, and of course there was no cure, only debilitating treatment with AZT. Duesberg went against considered opinion about the cause of AIDS in such major scientific outlets as *Cancer Research* (1987), *Science* (1988), the *Proceedings of the National Academy of Sciences* (1989), and *Pharmacological Therapeutics* (1992). His idea was that AIDS is caused not by HIV but by other things, like recreational drug use, AZT itself, and poor nutrition and sanitation.[39] If true, then AIDS would not be infectious, and programs aimed at reducing its transmission would be doomed to failure on the grounds that AIDS is just not transmitted.

Duesberg put forward his ideas, and they were firmly rejected by the scientific community. Duesberg and a small band of acolytes managed to raise their contrarian voices periodically, like Velikovsky and his supporters, and they were very good at raising criticisms about the normative ideas on AIDS but not very good at addressing the far more extensive criticisms of their own views. Finally, in 1993, AIDS researchers showed definitively in *Nature* that recreational drug use (not including injected drugs) actually had no relation at all to developing AIDS, as everybody but Duesberg and his crew already knew. Duesberg demanded the opportunity to publish a rebuttal. Being a high-ranking and honored scientist, he was accustomed to having his demands met. This time, though, it didn't happen.

Duesberg immediately claimed (like Velikovsky) that he was being denied his day in court, being silenced by the establishment. But this was a bit different. He was a part of the establishment, his ideas had been published by the establishment, his ideas had been taken seriously enough to have merited a test and a falsification. The silencing amounted to denying Duesberg space in *Nature* just to bash the study that had refuted his ideas. *Nature*'s editor, the redoubtable John Maddox, saw that as having rhetorical value, but not scientific value, and rejected Duesberg's rebuttal on those grounds.[40]

The issue is not that Duesberg's views had been denied an airing but that they had been aired and rejected—and the stubborn scientist just wouldn't accept it. That is a crucial distinction, and the failure to draw it can be a telltale sign of bogus science. This is not suppression; it is obsession. And it evokes another problematic image: "The fools! They called me mad!" Any scientist shaking his fist at the sky is simply not one to be taken seriously.

But if he ever actually *does* invent the plasmatic hypergravitational atomic death beam, look out!

7. Raising more difficult questions than it resolves

A scientific advance is one that satisfactorily answers a question that has been posed. Usually there are a lot of possible answers and diverse reasons for choosing one over another. Science aspires to choose answers that are supported by reliable empirical evidence. Even so, the evidence is often not so clear-cut.

An advance opens up new questions as it answers old questions. It wouldn't make much sense to unexplain things that have already been adequately explained and raise more difficult and fundamental questions than you have just answered. That's a good sign of a bad answer. Consequently, when evaluating the merits of a new idea, it helps to ask what intellectual baggage may be carried along with it. This information may reveal concealed implications of the idea or aspects of the quality of scientific thought that produced the idea.

In 1971, the scientific world was rocked by the discovery of skull ER-1470, by Bernard Ngeneo, a member of Richard Leakey's "hominid gang." As Leakey presented the fossil, it was 2.8 million years old, had a larger brain than anything else of that age, and had an almost vertical face, like a modern human's. The fossil seemed to say that virtually all the later human-like fossils, with more prognathic faces, were side branches, off the main line, which was already established 2.8 million years ago. Or else, as the paleontologist Basil Cooke thought, the skull was misdated by about a million years, and, as the anatomist Alan Walker thought, its face was inaccurately reconstructed.[41]

Quick—which was more likely: that the fossil was being interpreted poorly, or that everything we knew about human ancestry was wrong?

Framed in such a fashion, the answer seems ludicrously simple. Nevertheless, a decade-long battle over the importance of skull 1470 was ultimately resolved by forcing an acknowledgment that the fossil had indeed been misdated and misreconstructed. Too much bombast and not enough caution are never a successful recipe for the production of reliable scientific facts. More important, Leakey's interpretation of that fossil discovery directly implied that much of the intellectual framework already in place was wrong. That was not impossible, just a lot less likely than the prospect that the little bit of data that seemed to contradict it might be wrong instead.

Indeed, about the same time that skull 1470 was clogging the intellectual arteries of paleoanthropology, a new, broad theory of human evolution was being put forth by a Welsh writer named Elaine Morgan. In the "aquatic ape" theory, some seemingly disparate human features are explained by denying the basic story of a descent from the trees to bipedality and instead placing human ancestors in the water, wading and swimming.

In particular, this theory calls attention to the reduction of body hair and gain of subcutaneous fat deposits that differentiate us signally from the apes, then argues cleverly that these features are also found in other swimming and wading mammals, such as dolphins and seals. Consequently, these features may be regarded as indicative of a hidden crucial chapter in human evolution—an adaptation to the water. On first blush this may seem reasonable, but it does gloss over the fact that human skin is riddled with sweat glands for evaporative heat loss, and that human body composition is more dimorphic than body size—both of which are quite different from the situation in aquatic mammals.

Further, not only is sweating a useless adaptation in the water, but in the places of the body where we have retained hair most conspicuously (the armpits and crotch), the hairs retain smells from glandular secretions, for these are also the stinkiest parts of the body. That suggests an odiferous reason for their presence—a social signal, perhaps?—but one that would be useful only on land and perfectly useless in the water. Check out the armpits of a walrus sometime!

Another feature shared by aquatic mammals is the tendency of evolution to move their nostrils upward, to help prevent drowning. And yet our nostrils are in the same place as an ape's.[42] If anything, the nasal cartilage our species has evolved tends to orient our nostrils more downward than those of an ape, which would be positively counterproductive in the water.

So where does that leave the "aquatic ape" theory? The evidence in its favor amounts to a small and carefully selected set of traits, ignoring the other traits that might be expected to coevolve according to its central argument. Why would evolution latch on to subcutaneous fat but leave the nostrils alone? Why would we have evolved mechanisms to sweat copiously if our ancestors were aquatic? What good are olfactory signals, anyway—like those that emanate from the hairy spots—in the water?[43] The "aquatic ape" theory is forced to deal with these questions on a case-by-case basis, when they would seem to be integral to the theory itself.

This is not to say that the "aquatic ape" theory is wrong. Indeed, it is hard to know how we could even tell. It is to say, rather, that the theory answers questions that already have satisfactory answers (In what kind of environment did our basic adaptation arise? Why are we not as hirsute as the apes?) and un-answers questions that also already have satisfactory answers (Why did we develop apocrine glands in our axillary and anogenital regions? Why are our forelimbs and digits long and limber like those of landlubbing primates rather than short and stiff like those of aquatic mammals?). Once again, not the signature of a particularly useful scientific idea.

BAD, BOGUS, OR PSEUDO: IT'S ALL THE SAME

Everyone knows that astrology, UFOlogy, and faith healing are not science. Real scientists don't partake of them, and their absence says enough. To the extent that people are interested in horoscopes or ancient astronauts, they know that such ideas derive their legitimacy from other sources than do MRI scans and antibiotics. They are the comforting inanities of daily life.

But that isn't the problem of pseudoscience, for (aside from a few charlatans who dangle letters after their names or use beeping machines in their endeavors and actually misrepresent themselves as scientists), these domains are generally explicit in labeling themselves as alternatives to science.

At issue, rather, are the comforting inanities of *scientific* life, the nonsense that is said and done under the umbrella of science—and which makes it far more threatening, for it can indeed be misread for knowledge itself. For if scientists themselves don't distinguish it for us, how can the rest of us do it?

Ultimately, the racism of one molecular biologist, or the naïve genetic determinism of another, is worse than the racism or essentialized ideas about heredity held by the booboisie. Why? Because the scientists are supposed to be smarter. Their ideas are supposed to be that much more credible. Their blunders consequently end up giving science a bad name, something that Velikovsky, of course, was never actually able to achieve himself.

SEVEN Scientific Misconduct

The two most famous cases of fraud in the history of the life sciences are not at all prototypical. In the case of Piltdown Man, the deceiver was most likely not a professional scientist and thus had a motive other than personal gain for his work. In the case of the Midwife Toad, the deceiver—while never admitting his guilt—honorably blew his brains out upon the public revelation that his work had been falsified rather than face the humiliation. Far more often than not, scientific fraud is committed by scientists themselves (hence the name, *scientific fraud*), and they hardly ever acknowledge wrongdoing.[1]

The reason they don't is that they have an important professional incentive not to do so. That incentive is encoded in a normative operating premise of science: it is acceptable to be incompetent, but not acceptable to be dishonest.[2]

Using cutting-edge technology to extract and analyze DNA from chimpanzees in the field, Pascal Gagneux and his colleagues published a fascinating discovery in 1997. Over half of the infants in the social group they studied at Taï, in West Africa, had apparently been fathered by nonresident males—since they couldn't match them up to the DNA of any of the adult males in the group. This was a widely publicized result, but there were intricacies involved at several steps of the analysis. A lot of things could have gone wrong, and unfortunately they had indeed gone wrong. A similar study from Gombe in East Africa found that all the offspring were indeed sired by resident males, and a reanalysis of the Taï chimps found that they too had been overwhelmingly sired by resident males. Gagneux et al. took it all back in 2001, admitting they had "inaccurately genotyped" the chimps for reasons ranging from the technical ("allelic dropout") to the amateurish ("sample mix-up"). D'oh!

No one ever accused them of misconduct; we just have to hope that their other work is a heck of a lot more appropriately cautious, conscientious, and competent.[3]

Jeez, we all make mistakes, after all. But only a few of us are cheaters. And although the ultimate consequences of errors and deceptions are the same (bullshit passing for reality and retarding the progress of science), the consequences for the individual scientist are quite different. Caught in an error, the scientist may have to eat a bit of humble pie and may even be transiently ridiculed or scorned (or not; see Lord Zuckerman, chapter 6). But caught in a lie, a scientist has violated the public trust, and that is a problem, especially if the public has been funding the researcher.

The result of this concurrent asymmetry between the professional consequences of fraud and error, despite the symmetry of their results for the scientific endeavor, is that it sets up a convenient plea bargain. Any scientist accused of fraud can mount a defense by saying it was "just" a mistake. Being found guilty of mistakes—however idiotic or egregious— can nevertheless permit a scientist to remain in business, on the presumption that errors are aberrations, and there but for the grace of God go the accusers themselves.

There are profound implications, however, for the integrity of the scientific literature. Take, for example, the Imanishi-Kari case (below), in which investigators looking for dishonesty in the crucial paper concluded that they couldn't identify any—but that they *could* identify numerous "errors" of varying degrees of importance. The problem is that they ostensibly weren't looking for error, they were looking for fraud. The paper was selected for scrutiny on account of its possibly dishonest content, not for any possible mistakes. So, a paper published in a major journal in the field, chosen at random with respect to the possibility of mistakes, turns out to be rife with them. What does that mean about the quality of the published literature and the effectiveness of the review process?

Nobody really knows. But when people have tried to hazard an estimate, the results have been discouraging. In 1981, the cardiological research carried out at Harvard and Emory by John Darsee was discovered to have been falsified. The compromised output ran to 109 published papers with forty-seven coauthors, none of whom had been accused or suspected of anything untoward. Two NIH researchers interested in scientific fraud—and nearly alone in that interest—subjected those papers to a fine-grained analysis and found them to be riddled with errors, contradictions, and misstatements, *even though they had not been selected for those characteristics;* they had been selected because a coauthor was dishonest. If peer review is what guarantees the integrity of the scientific literature, they argued, it isn't working very well. If there are so many errors in the papers, why didn't the coauthors catch them? And why didn't the reviewers catch them? Perhaps if coauthors were more scrupulous about checking what they have signed on to, fraud might also be caught more readily.[4]

Fraud, after all, was the subject. What does it mean to be able to defend a scientist's reputation by saying their work was not dishonest but "merely" careless and inept? And what does it mean to be able to find such a high level of mere carelessness and ineptitude in a sample of publications by major researchers at major institutions, published in major journals?

The consequences of plea bargaining a fraud accusation "down" to a confession of error are salutary for both parties. The accused may be spared their reputations, and the accusers are spared having to prove an intent to deceive—the *mens rea,* or "guilty mind." That sort of thing is almost impossible to prove, anyway.

SCIENCE AND HONOR

Fraud is interesting precisely because it goes against the accepted norms. Nobody believes that you ought to fudge or fake your work, and certainly nobody teaches that to their students. They don't teach that in the government either, but we have come cynically to expect some nontrivial level of corruption there. Why shouldn't we expect some from science?

According to the social historian Steven Shapin, the answer seems to be that science emerged in the seventeenth century as a new philosophy—but not in isolation from other cultural forms. In addition to being a revolutionary approach to thinking about nature, it was associated with specific political and social institutions and developed its own standards of truth and trust.[5] That trust was founded on the gentlemen's code of honor, associated with the social and political status of its practitioners. As long as science was the leisure pursuit of the idle wealthy class, the code was effective; a betrayal of trust would be unthinkable, like early Christian martyrs turning one another in to the Romans.

As science increasingly became professionalized in the late 1800s—that is to say, as science became more of a job and less of a hobby—the gentlemen's code assumed diminishing relevance. When the currency of competition was relatively simple—priority of discovery—there was not much to fudge. But as the currency became more complex—prestigious academic posts, grants, patronage, a livelihood—the stakes went up. The question "How do I know what or whom to believe?" is a question as old as the evolution of language itself. And the reliance that we place on those who provide the answers—your mother, your shaman, your priest, your doctor, your president—encodes the complexities of social enculturation and maturation.

Science is a cultural form in which truth is likewise maintained by trust; surely your parents or your imam wouldn't lie to you, and neither would your scientist! And yet it was not pleasant to learn the truth about Santa Claus; and religious apostates are made, not born. So what happens when we learn that we can't always believe what our scientists tell us, either?

Presumably we would want some degree of confidence that a dishonest scientist would be caught and punished—much as a dishonest lawyer would be disbarred, or a dishonest priest defrocked. That result, however, requires some acknowledgment that there is a problem of honesty as well as the means for defining it, rooting it out, and dealing with it. This produces the first strange situation about scientific fraud: there is no Science Police.

Just as lies are as old as language, scientific fraud is as old as science. And although the early gentleman-scholars developed an informal moral code decrying it, there are nevertheless some famous stories about the early days. Galileo sometimes blurred the distinction between experiments he did and experiments he could have done. Newton's mathematical brilliance helped him claim a degree of precision for his results that was far greater than any he could actually obtain. When confronted with the appalling possibility that Leibniz might actually have invented calculus independently of him, Newton formed an impartial committee to investigate the claim and then wrote its report himself, rejecting Leibniz's claim. And need we even ask whether Brother Mendel's quantitative demonstration of the 3:1 ratio of hybrid pea plant crosses came *before* or *after* his expectation of the ratio itself?

Historians have long pondered these and other acts of the early scientists, but perhaps the 1726 case of Professor Johann Beringer of the University of Würzburg is the first well-known case of actual fraud. Despising Beringer for bombastic pedantry, two of his faculty colleagues contrived to dupe him with fake fossils. And the old prof bought it—hook, line, and sinker—and made plans to publish his great fossil trove, for all to ponder. Having made their point, the hoaxers tried to talk him out of publishing it, by planting ever more ridiculous fossils—including ones with ancient sacred inscriptions. Undaunted, Beringer published his great work and subsequently demanded (and received) a formal inquiry into the identities of the hoaxers. One of Beringer's student assistants fingered the other faculty members. Beringer then spent the rest of his life so ashamed that he tried to buy back all the copies that had been sold.

Beringer is remembered as an old fool who let his expectations shape his interpretations and was punished accordingly, as if a hero in a Greek tragedy. But at least he fared better than the hoaxers: one died shortly thereafter, and the other was exiled.[6]

The Beringer story is not usually considered a case of scientific fraud; that is to say, it is usually recounted as a fable about gullibility, not about moral conduct. But what of those two malicious science professors who falsified the fossils? Are we supposed to take their sociopathy for granted? And what of their own work—are we to assume that someone who would falsify other people's data would not falsify their own?

Those fellows didn't actually leave us any work, so the question is strictly academic. But it raises a crucial issue about the establishment of trust in the scientific community and its breach. Once someone is a falsifier, does it mean that the one detected case is the only one? Or is it more likely that everything that person has ever done is necessarily compromised? There is, obviously, another plea bargain here: "Okay, this particular piece of work is dirty; but all the rest I'm publishing are clean." But the opposing viewpoint is also compelling: that scientific integrity is like virginity—once you lose it, you can never get it back.

In the Beringer case, the nature of the falsifiers' own work doesn't matter. Somehow the system worked, even though, paradoxically enough, there was no system.

As early as 1830, the problem of honesty in science, and the failure of its effective regulation, was taken up by the mathematician Charles Babbage. He identified four kinds of fraud that distinguished the true scientist from "pretenders." These were hoaxing (in which someone else's data are falsified); forging (falsifying one's own data); trimming (disregarding outlying data or negative results); and cooking (making unconvincing data into convincing data).[7] In the age before the professionalization of science, there was obviously a lot of that going on; to a large extent the credentialing process that would accompany the professionalization of a science helped to build a wall between the scientist and the "pretender." This was certainly the case in archaeology, bedeviled by charlatans and hoaxers in its early years, as it eventually produced a knowledgeable class of professional scholars bound by the honor code of science.[8]

PILTDOWN MAN

The highest-profile fraud cases are not necessarily the biggest or the most important, but they are the ones that get people interested in the subject, and each has been instructive.

History is of course political, and no less political is the universal history of our species. In the post-Darwinian world, the record of human ancestry would be read as a set of natural facts, preserved in the geological column as fossils. By the early twentieth century, brutish archaic human-like remains had been found in various countries of northern Europe—France, Belgium, Germany—and even as far away as Java. The oldest human remains that had been found in England, however, were unmistakably modern in form; the greatest intellectual, military, and economic power in the world was but a footnote to the glorious ascent of the human species. But that all changed in 1912, when an English lawyer and naturalist named Charles Dawson brought some cranial fragments to the attention of his friend, the paleontologist at the British Museum of Natural History, Arthur Smith Woodward.

Dawson and Woodward began to conduct excavations at Dawson's site, the gravels at Piltdown, and soon turned up an array of fossils. In December 1912, they revealed their results, the discovery and reconstruction of the noble human ancestor—*Eoanthropus dawsoni,* or Dawson's Dawn Man. It seemed to have teeth resembling an ape's (suggesting brute strength?) and a brain approximating a man's (suggesting nascent intelligence?)—in short, a venerable ancestor for the human race. Almost immediately questions were raised about the association of the ape teeth with the human brain, as well as about the reconstruction. Dawson and Woodward answered with more finds from the site: archaic tools, more apish teeth and human skull bits, and the fragmentary remains of other extinct species. After Dawson's death in 1916, no other fossils were ever found there, but these were enough. Piltdown Man was the mythic ancestor; scientific reputations would be made from him, and his bones would be kept like medieval holy relics, with copies of them made available to other scholars.

In a magazine article published in 1914, the American paleontologist William King Gregory repeated a rumor he had heard while examining the remains: "It has been suspected by some that geologically they are not old at all; that they may even represent a deliberate hoax, a negro or Australian skull and a broken ape-jaw, artificially fossilized and planted in the gravel-bed, to fool the scientists."[9] But he rejected that suggestion, and no similar suspicion ever entered the public record.

By the 1940s, however, Piltdown Man's anatomy was enigmatic. Other fossils from Africa and Asia showed that the human teeth had evolved before the human brain, for they combined human-like teeth with a primitive small brain. How could that sequence be different in England? Some non-British scholars, such as Marcellin Boule and Franz Weidenreich, had been casually or coyly dismissing Piltdown Man for years. Most others felt that it was "part of the canon" and had to be at least mentioned.

Finally, in the early 1950s, a new generation of scholars—Joseph Weiner, Kenneth Oakley, and Wilfrid E. Le Gros Clark—subjected the Piltdown remains to renewed and hi-tech examination. Even at low-tech, they could now see that the canine tooth had been painted brown, the molar teeth had been filed down, and most of the other materials had been stained to look older than they really were. The skull was from a human, but the jaw was from an orangutan. Even the tools had been planted; one had even been thought to resemble a primitive cricket bat. Nothing was as it had originally seemed; it was all a fraud.[10]

The question that has dominated Piltdown Man over the ensuing half-century is, Who did it? The list of motivated suspects is impressive. We have the haughty professors, their academic rivals, and their underpaid and unappreciated staff. We have practical jokers, venomous colleagues, and even a lawyer. Finally, we have a local celebrity author—Sir Arthur Conan Doyle—who was friendly with Charles Dawson and published his own novel *The Lost World* in serial form just as Piltdown Man was coming to light. The novel just happened to include a fight between professors and ape-men and an underlying theme of how to convince skeptics of a wild story in the face of scanty evidence.

But that misses the point. It doesn't really matter who did it; what matters is how it worked. The clever bastard's identity is far less important than the social processes that permitted him (by the way, all the suspects are male) to get away with it. The perpetrator is long dead; but those circumstances may still exist and consequently may conceivably still pose a threat. And Piltdown Man does indeed illustrate some key social features of scientific fraud.

First, without someone to press the case for fraud, it is unlikely to be discovered in a timely fashion—because it is not likely to be suspected in the first place. The debate over Piltdown Man was over whether the association between the ape jaw and human cranium was real or accidental. To the extent that a third possibility was voiced—that it might be neither real nor accidental—that possibility was voiced privately and quietly. There was no reason to look for fraud because there was no one saying (audibly) that it might possibly be fraudulent.

Second, the further removed one gets from the primary data, the more unlikely one is to detect fraud. Once again, the scholars who controlled Piltdown Man were generous with fossil casts, but few got to see the real thing. The American physical anthropologist Aleš Hrdlička was one of those who did get to see it, and he left puzzled by what seemed to be wear patterns on the molar teeth that indicated heavy side-to-side chewing, which would be anomalous for an animal with large, interlocking canine teeth. Perhaps if he had looked more closely, or if others had examined the originals, the fact that the molar teeth had actually been filed down might have been recognized earlier.

And third, nothing inflates dishonest work like power. In the case of Piltdown Man, we have not only the nationalistic ideological power of "the first Englishman" but the social power of the academic hierarchy that was heavily invested in him. Three of the major students of Piltdown Man were knighted in the 1920s: Arthur Keith, Grafton Elliot Smith, and Arthur Smith Woodward. A fourth, E. Ray Lankester, had already been knighted by the time he participated in the Piltdown Man analysis. You'd have to be crazy to go up against institutional clout like that with a charge that their favorite fossil was fraudulent. Moreover, since they were, after all, the greatest and most prestigious experts, other scholars tended to rely somewhat lazily on secondhand pronouncements about Piltdown Man rather than on their own critical faculties. That laziness, however, is the flip side of one of the primary expectations of science: you have to be able to rely on the judgments of the experts. If you can't, and you must call into question everything everybody says, then science simply cannot function.

THE MIDWIFE TOAD

In the second most famous case of fraud in the life sciences, we find again the compromising influence of political ideology, of a different sort. In the late nineteenth century, two great sociopolitical views were playing out their final battle. One position held that differences of class—wealth and clout—were the result of constitutionally based differences between the people composing the classes. This view encoded a justification for the hereditary aristocracy that had dominated European life for centuries. The other position held that there was no natural basis for that aristocracy; society could run as well, perhaps better, without inherited class divisions. In this view, anyone had the natural potential to attain wealth and power; what distinguished the haves from the have-nots was not so much their respective familial traits as the circumstances of birth and opportunities open to them.[11]

Theories of heredity intersected with these political views. On one side, the natural endowments of the aristocrats were taken to be self-evident and self-perpetuating. On the other side, the endowments of the workers were taken to be at least potentially as good as those of the rulers; under appropriate circumstances, their own qualities could be modified into those of the rulers. On the one side, natural divisions predestined social divisions; on the other, people could be transformed through time into other kinds of social beings.

It should not be terribly surprising, then, that social progressives like Karl Marx found much to admire in the "transformism" of Darwinian biology. A generation later, however, the German biologist August Weismann was promoting something he called "neo-Darwinism"—which held that nothing that transpires in an organism during its life can be stably transmitted to its offspring. Heredity is stable; the future's "germ-plasm" is the same as the present's "germ-plasm."[12] Significantly, though, this separation of a stable repository of reproductive cells from the developing and changing body, and the next generation's constitution from the former rather than from the latter—what Weismann called "the continuity of the germ-plasm"—could be seen as a conservative antidote to the political views that saw continuous progressive change as the leitmotif of life on earth, including human existence.

The rise of Mendelian genetics in 1900 seemed to cement this (paradoxically non-Darwinian) "neo-Darwinian" theory. Hereditary units were transmitted immutably across generations, readily entering into new combinations but only rarely changing themselves. That narrative is, however, a considerable oversimplification, a product of the Whig or presentist version of scientific history that came to be told decades later, after the "evolutionary synthesis" of the 1940s.

In fact, biologists of the early twentieth century were quite divided on the reality of the inheritance of acquired characteristics. These rival biological theories coexisted tensely—and to a large extent still do—particularly over their relevance to human affairs. Just how much social change is permitted by human biology? Are class differences innate, the products of stable and inequitable natures? Or are they ephemeral and mutable, the products of human agency and persistent human evil? The neo-Darwinian dogma, dictated by the stability of alternative alleles across generations, seemed particularly compatible with the first alternative. Early Mendelians, especially in the United States, infamously envisioned genes for intelligence/stupidity that were conceptually modeled on the dichotomous green/yellow and wrinkled/round traits in peas. You either had it and inherited it, or you didn't—but you couldn't earn it.

With the aid of decades of hindsight, the connections among these ideas seem rather tenuous now, but the early 1900s was indeed a period of considerable ferment in genetics. Most biologists were unsure whether evolution was goal directed, whether acquired characteristics were inherited, whether natural selection was the primary force producing adaptation, and whether cultural progress was tightly tethered to human biology.

And all of these issues were eclipsed by the really important things going on in the world—the Communist revolution in Russia and the world war. The revolution could be seen (optimistically) as an example of social progress—the final collapse of ancient aristocracies and their replacement by newer (and fairer) social forms. You had to look a lot harder to see anything about World War I optimistically, except its end—which still left most of Europe in far worse shape than at its beginning.

So what did the new biology have to say about human society, social progress, and the future of the human race? The inheritance of acquired characteristics seemed to offer a rosier future of possibilities than the neo-Darwinian straitjacket of August Weismann, Gregor Mendel, and their followers—Darwin's work itself being largely tangential to any of these debates.

And therein lay the interests of a charismatic Austrian biologist named Paul Kammerer. It was the era of a new experimental zoology and of the study of basic genetic processes in simple, rapidly reproducing model organisms, like Herbert Spencer Jennings's paramecium and Thomas Hunt Morgan's fruit flies. Kammerer, however, worked with more complex and slower-developing model organisms—amphibians. By 1910, he had obtained interesting results of diverse kinds. He induced salamanders to change the color of their stripes, eyeless newts to develop functioning eyes, and other species to develop other characteristics, which were stably passed on to their progeny.

His successes were controversial from the start, however. Weismann referred to him privately as "a little, miserable, sticky Jew, who has proven himself on earlier occasions to be a quite unreliable worker"[13]—which indicates the complexities of his scholarly and social relations. In fact, there were persistent complaints about the quality of documentation Kammerer provided in his papers. To be fair, these were the early days of photographic reproduction, and it was common practice for scientists either to draw pictures of their results or to retouch photographs to make the critical parts of the image more obvious.[14] Kammerer, however, was making strong claims, which required high documentary standards, and he was simply not up to the task. As described by his contemporary, the German Jewish expatriate geneticist Richard Goldschmidt, Kammerer was "a very highstrung, decadent but brilliant man who spent his nights, after a day in the laboratory, composing symphonies" but "simply did not know what an experiment amounted to."[15]

Kammerer's most celebrated experimental result involved *Alytes obstetricans*, known as the common midwife toad. Since most toads mate in the water, the males develop dark thickenings called "nuptial pads" on their forelimbs, which allow them to maintain physical contact with the female while mating. This species, however, is a land dweller, and its males neither need nor develop these nuptial pads. Kammerer forced them to mate in the water, got the males to develop the nuptial pads, and got them to transmit these nuptial pads to their sons, even on dry land.[16]

Other biologists were frustrated with Kammerer's glibness about his results and his lack of rigor in demonstrating them before World War I, but after the war the situation only intensified. Much of his work had been disrupted, if not destroyed, by the war, and by the early 1920s Kammerer was also in difficult financial straits. He was nevertheless still a handsome, cultured intellectual and a good self-promoter. He resigned his laboratory post in 1921 to write and lecture full-time.

Matters came to a head in 1923, when Kammerer embarked on a protracted lecture tour. The feisty geneticist William Bateson—who had himself coined the term *genetics* in 1906—had been frustrated for years in his attempts actually to examine Kammerer's midwife toads at first hand. Bateson now publicly challenged Kammerer in the pages of *Nature,* and Cambridge students and faculty took up a collection to bring Kammerer out to lecture. He spoke and showed his lantern slides as well as his last, old specimen—which seemed to have something dark on the tips of its forelimbs—but Bateson would have none of it and continued to press the issue in *Nature.* Other Cambridge biologists were significantly more sanguine about Kammerer's work. "Kammerer begins where Darwin left off," said one. "He has made perhaps the greatest biological discovery of the century," said another.[17]

Once again, however, Kammerer drew his own interpretations about human society, not just about froggies. "The results make it probable," he said, "our descendants will learn more quickly what we know well, will execute more easily what we have accomplished with great effort, will be able to withstand what injured us almost to the point of death. Where we sought they will find. Where we began they will accomplish."[18] At face value, these conclusions were, and are, probably true; but they are derivable from cultural history, not from amphibian mating habits.

In New York in December 1923, Kammerer was received warmly by some prominent biologists, coolly by others, and continued to speak to the press of "acquired characteristics, which, I believe, will play a supreme part in the future evolution of the human race." "I would suggest, first, that it be used to eliminate race hatred. . . . Civilization and humanity, as a whole, must derive the full benefits of what has been discovered about acquired characteristics of heredity."[19]

Can't really argue with that.

Back home in his native Vienna, Kammerer grew increasingly frustrated by the lack of appreciation for his work and ideas shown by his colleagues. William Bateson died early in 1926, and while that might have been a godsend for Kammerer, it was really the beginning of the end for him. Later in the year, the herpetology curator at the American Museum of Natural History, bearing the impeccably regal name of G. Kingsley Noble, visited Kammerer in Vienna and got a close look at the midwife toad. Kammerer, meanwhile, was being courted by the Soviet government, which offered him the directorship of a laboratory at Moscow University. Reluctant to leave Vienna, Kammerer nevertheless accepted the offer and made plans to move to Moscow. Then *Nature* published the results of G. Kingsley Noble's analysis of the nuptial pads of Kammerer's midwife toad: there were no nuptial pads; the darkened patches were the result of injections of India ink.

On September 22, 1926, Kammerer wrote his letter of resignation to the Moscow Academy of Sciences.[20] The next day he killed himself.

What lessons can be drawn from "the case of the midwife toad"?[21]

First, unlike Piltdown Man, about which rumors also traveled along the grapevine, there was a scientist willing to voice his doubts publicly and to press the work in question to meet a burden of proof. William Bateson acted in the role of nemesis, an opponent or challenger of the work—as opposed to a whistle-blower, or insider, as in the Summerlin and Imanishi-Kari cases, below.[22] Both the nemesis and the whistleblower expose themselves to considerable personal and professional harm, which is why the more anonymous but largely ineffectual grapevine is the favored venue for voicing one's suspicions. But the grapevine also favors those with access; senior scholars can use it to discredit junior scholars far more readily than the other way around. Moreover, with reputations and livelihoods on the line, a nemesis risks a reciprocal charge of libel and therefore has to be very certain of the case before even coming forward with the charge.[23]

This exposes a crucial difference between the scientific and legal communities. In law the accused is innocent until proven guilty, and the case is built and presented after the initial charge is filed. In science, however, the burden of proof must fall squarely first upon the original investigator. The trust that goes into scientific publication includes an understanding that all of the information cannot possibly be included in a scholarly report but is digested into its relevant components in a publication; if needed, it all will be made available. Kammerer had the habit of not documenting his claims adequately for the skeptical, only for the already convinced. The purpose of a scientific report, however, is to convince the skeptics, not the choir; that is what initially made it different from other forms of knowledge.

Without a Science Police, however, a nemesis must act largely alone, building and presenting a prosecution against the original work without the benefit of subpoenas or the state apparatus of a legal case. There are only three reasons why researchers would not release the data behind a published paper: there are no data; they are reluctant to help a competitor; or they are afraid of what someone else will see in the data. None of them is considered valid justification. And although we might hope that journal editors would demand the availability of data, at least to ensure the integrity of their own publication, the fact is that they are under no supervision either. As late as 1988, when pressed by other scholars for documentation of questionable work that had been published in the *Journal of Molecular Evolution,* a senior researcher (who was also an associate editor of the journal) responded, "We are not compelled to give our data to just anyone who asks," and was actually supported by the editor in chief.[24] In which case, what's a nemesis to do?

A second issue raised by the Kammerer affair actually arose a few decades later. The Soviet Union, it turns out, never quite gave up on Kammerer. In 1929, their commissar of education wrote the screenplay for a film called *The Salamander,* loosely based on Kammerer's life but with the scientist being the victim of a frame-up and saved from pulling the fatal trigger at the last moment by his devoted love, played by the commissar's wife. The film was sensitive enough to be banned in pre-Nazi Germany.[25] And as long as Kammerer's papers and books were still "out there," there was nothing to prevent subsequent polemicists, popularizers, or scientists from citing it—either with or without the knowledge that the work had signally failed to pass the test of reliability.

Indeed, by the 1950s, that is exactly what was happening, much to the amazement of the geneticist Conway Zirkle. Zirkle raised a question that the scientific community has subsequently had to grapple with as more research is discovered to be compromised: How do you deal with derivative work that still cites fraudulent primary work as if it were actually reliable? Or, more fundamentally, what do you do with work that has been shown or admitted to be unreliable, so that later workers aren't misled by it?[26]

There are a number of relevant variables: how old the work is; whether it has been formally retracted; whether it has been shown to be dishonest or "merely" incorrect; and the manner in which the literature is accessed. Electronic journals have more flexibility than their printed versions, which are likely to be bound and up on library shelves. And certainly science librarians have better things to do than look up each article that has been retracted, intercalate a note to that effect (assuming someone has specifically apprised them of it), and rebind the volume. Even so, there are limits to what electronically accessible journals can do or are willing to do.

If you access the journal *Science* online and look up, for example, the paper "An Organic Solid State Injection Laser" by Jan Hendrik Schön et al., published in 2000, you will find appended to it the retraction published in 2002. Schön, a prolific and brilliant nanotechnology physicist who was once considered to be on the fast track to the Nobel Prize, was found to be falsifying his work at Bell Labs.[27] If you access the journal *Nature* online and look up the 1997 paper "Furtive Mating in Female Chimpanzees" by Pascal Gagneux et al. (see above), you will find the paper intact. So, too, is the 2001 retraction (for ineptitude rather than dishonesty)—but they are in separate places, disconnected from each other. Worse still, if you access *Science* and look up Richard Wrangham's 1997 essay "Subtle, Secret Female Chimpanzees," which summarized and discussed that *Nature* paper, you will encounter no indication that the work under discussion is no longer regarded as reliable and was in fact retracted.

If these situations seem complicated, consider a simple solution to the problem from the journal *Human Immunology.* A paper on the genetic relationships of Jews and Palestinians was published in its issue of September 2001, written by the issue's guest editor, a native speaker and writer of Spanish. After complaints that some of the phrasing in the paper seemed intended to delegitimize the Israeli presence in the Middle East and some threats of resignation from the board of the American Society for Histocompatibility and Immunogenetics (tensions were high, for this was the very month of the 9/11 attacks), the journal decided on a bizarre course of action. It retracted the paper for reasons of neither breach of honesty nor lapse of scientific competence but for "historical errors and inconsistencies, inadequate and misleading references and inaccurate maps" and then wrote to libraries asking them to excise the pages from their copies of the journal.

In the first place, it seems ridiculous to expect that a paper on the genetics of Israelis and Palestinians could possibly be free or independent of the geopolitical turmoil that has existed between its subjects for decades. (Why, after all, would anyone select these particular groups for genetic comparison in the first place?) But to ask librarians then to mutilate their books is simply nutty; it's not the kind of thing librarians are at all inclined to do (except perhaps Nazi librarians). If you try to access it online, you get this message: "Article has been withdrawn by the American Society for Histocompatibility and Immunogenetics (the copyright owner), the Editor and Publisher, and will not be available in electronic format." To actually read the article, then, for whatever your interest in it may be, you will have to rely entirely upon the professional scruples of your local librarian or find it elsewhere on the Web.[28]

OF MICE AND PEN

It is widely held that the modern era of public interest in scientific fraud began with the 1974 scandal at Memorial Sloan-Kettering Cancer Center in New York.[29] There, a top-rated senior immunologist named Robert Good had been recruited from Minnesota to direct the center and brought with him a dermatologist protégé named William Summerlin. Summerlin first got his name in the *New York Times* on March 30, 1973, (even if misspelled as Sommerlin) under the heading "Lab discovery may aid transplants." (It was later found that he had made the local paper in Atlanta some years earlier, after being accused of cheating on an exam in medical school.)

Summerlin was now head of a laboratory, was working on a method to facilitate skin grafts from person to person, and "said that several other laboratories [had] duplicated his findings." His work was mentioned favorably again a few months later, with the ominous qualifier that "some doctors challenge this finding, but if it is confirmed . . ."

But researchers in Cincinnati and London were frustrated, first by their inability to get the same results as Summerlin, and second by their inability to determine precisely what they were doing differently, for Summerlin was very sparing on the details of his procedures. Even a post-doc in the same lab, trying to replicate the results, couldn't do it. On March 26, 1974, Summerlin had an early-morning meeting with his senior colleagues to convince them of his transplantation results. He showed them two white mice, each with a patch of black skin on its back. They were convinced.

Summerlin returned the mice to their cages after the successful forty-five-minute meeting and turned them over to the lab technician. The technician, however, knew these mice and didn't remember such prominent black patches on them. Curious, he applied a bit of alcohol to the black spots, and they came right off.

That technician told the head technician what he had just found, who told a visiting researcher, who told a post-doc, who then called the boss—Dr. Good—out of another meeting and told him what they had all just learned. Good suspended Summerlin immediately and initiated a comprehensive investigation. Two months later, the results of the investigation were made public: Summerlin had drawn the black spots on the mice with a felt-tip pen on his way to the March 26 meeting. He had also promoted gross falsehoods concerning the results of some other experiments involving rabbits, and even his old work at Minnesota seemed to be characterized by a "lack of properly organized and analyzable data."

Sloan-Kettering's report actually avoided the eff word—*fraud*—instead judging Summerlin guilty of "misrepresentation." Summerlin admitted drawing on the mice but maintained that "it was all a misunderstanding." Lewis Thomas, president of the Sloan-Kettering Center, decided that Summerlin had temporarily flipped out and gave him a year off, with pay. Summerlin, with no shortage of chutzpah, held his own press conference and blamed his mentor, Dr. Good, for placing him under excessive pressure. Thirty years later, he was believed to be practicing medicine in rural Louisiana.[30]

The first thing that jumps out at us from the Summerlin case is the extraordinary serendipity with which the fraud was actually detected. Several of the principals were incredulous, saying things like "Why would anyone falsify data? You can't get away with it!" *Science* wrote it up as "a story without a hero," but of course there was a hero—the mousekeeper, a technician named James Martin, who assumed the role of whistle-blower. Had Summerlin himself wiped off the ink before returning the mice, the fraud would indeed have gone unnoticed. And if he had subsequently gone on to another research project, no one would ever have been the wiser. Or, if James Martin—who surely made but a fraction of Summerlin's salary—had (a) not noticed anything odd about the mice, or (b) gone instead to Summerlin as an extortionist, or (c) just not given a shit, any of which is conceivable, the fraud would have gone undetected. Moreover, given the route the incriminating information took up the Sloan-Kettering hierarchy, if any of the people in the loop had (a) been loyal to Summerlin, or (b) not believed James Martin, or (c) not cared themselves, the fraud would have gone undetected. It would have been, at worst, a weird and anomalous result, which might well have been perpetuated for as long as Summerlin cared to. For the fraud to be discovered, there had to be concern enough about the work to merit scrutiny, care enough about the work to merit reporting the discovery of the ink, and someone with enough seniority and integrity to do something about it. Another administrator might simply have made a higher priority of squelching it.

Indeed, that was the first instinct of the laboratory chief in a high-profile case that happened at Yale just a few years later, under circumstances even more serendipitously weird. A researcher at NIH named Helena Wachslicht-Rodbard submitted a research article on hormone levels in anorexic women to the *New England Journal of Medicine* in 1978. The journal sent it out for review to three anonymous referees, among them a high-ranking endocrinologist at Yale named Philip Felig. Felig agreed to review it but simply passed it on to his protégé, Dr. Vijay Soman. Soman, who was working on a similar project, promptly plagiarized the ideas and some of the sentences in the manuscript and told his boss to recommend rejecting it, which the boss did. Then Soman sent his own newly augmented manuscript (coauthored with his boss) to the *American Journal of Medicine*. The *American Journal of Medicine* sent it out to a distinguished researcher at NIH named Jesse Roth, who agreed to review it but instead also just passed it on to his own protégé, who was Helena Wachslicht-Rodbard.

Imagine her surprise at finding parts of her own work submitted to another journal under someone else's names! Assuming the role of nemesis, she pressed the case for thievery but was told to lay off. The two bosses were old friends and would settle the matter quietly. Unsatisfied, she continued to press the case, ultimately forcing a three-way confrontation involving the thief (Soman), the chief (Felig, who was being recruited for a prestigious chair at Columbia), and their dean. After two years and an intensive investigation that nobody but the nemesis wanted, it finally emerged that very little Soman had published, with Felig's name on it, could actually be substantiated. Both of the junior people ended up leaving science—one in disgrace and one in disgust.[31]

Soman, like Summerlin, knew the odds were greatly in his favor. Both men were just a tad careless, but, most important, they were just catastrophically unlucky.[32]

The second thing about the Summerlin case is that the timely exposure of the data falsification again turned crucially on the knowledge and actions of a whistle-blower or nemesis—as with William Bateson for the toads and, in this case, James Martin for the mice. Without their involvement, both might well have turned out as Piltdown Man did—an odd result, some whispered allegations, but ultimately a successful con over the career, if not the life span, of the perpetrator.

As a result of the cluster of fraud cases in the late 1970s—Summerlin, Soman, Darsee, and others—some government interest in the integrity of the scientific literature began to be raised. Federal hearings in 1981 brought out spokesmen for science to assure Congress that fraud was rare and pathological, although they had no data to support the point. The editor of *Science* avowed that "99.9999 percent of papers are accurate and truthful," while a distinguished physicist condescendingly called attention to "the indelible line that separates harmless fudging from real fraud," although without explicitly saying whether the work containing all that "harmless fudging" should still be considered either accurate or truthful. At best, that would mean that scientists use very subjective and counterintuitive criteria for gauging accuracy and truthfulness; at worst, it would mean that a lot more scientific work is inaccurate and untruthful.[33]

When data began to be collected, they indicated that fraud is not so much a rare pathology as it is a nonnormative but pervasive aspect of the society in which it occurs—like marital infidelity, farting loudly at the dinner table, or parking in a handicap space. It's disreputable, and you surely wouldn't do it yourself, but you know people who do, or who have. But that is not what scientific fraud is supposed to be like. It's supposed to be like incest or cannibalism—bizarre, exceptional, disgusting, and deplorable—as the scientists represented it to Congress and to the public. And if there really is a lot of "harmless fudging," then what criteria explicitly distinguish that from harmful fraud?

A survey found that close to 90 percent of undergraduates in introductory biology courses admitted to manipulating lab data "almost always" or "often" in order to get a better grade. A chemist dismissed the relevance of these data for understanding the behavior of scientists, clearly not appreciating that scientists begin their academic careers sitting in those very classes.[34] And that leaves open the important question: as scientists proceed up the professional hierarchy, do they tend to get more honest and transparent in their work, or do they get more clandestine and dishonest in their reportage?

When the American Association for the Advancement of Science took a survey in 1991, it found that over a quarter of the 124 respondents had firsthand familiarity with a fraud case but that very few thought it had been satisfactorily resolved. A 1993 study published in *American Scientist* (after being rejected by *Science*) found that (with a 65 percent response rate) between 5 and 10 percent of scientists surveyed believed they knew a faculty member who had falsified research results. While that is a reasonably small proportion, it is nevertheless five orders of magnitude higher than the editor of *Science* had claimed a few years earlier. And that was with a narrow definition of misconduct (see below). When the definition was broadened, close to half had some firsthand acquaintance with it.[35]

In 2005, *Nature* published the results of a larger survey in which scientists anonymously discussed their own behaviors. While fewer than 1 percent admitted to "falsifying or 'cooking' research data," more than 15 percent acknowledged "dropping observations or data points from analyses based on a gut feeling that they were inaccurate."[36] Those gut feelings, of course, have the power to transform an (unpublishable) ambiguous result into a (publishable) statistically significant one.

This in turn raises the question of whether there is continuity or discontinuity between dropping a deviant data point based on a gut feeling that it's wrong and coloring in a patch of mouse skin based on a gut feeling that the mouse skin should be that color anyway. Is what Summerlin did a bit worse than what scientists ordinarily do, or is it simply different from what scientists ordinarily do?

The only way to grapple with that question is to define scientific misconduct rigorously, to establish what lies within its domain. But even that is hotly contested. If we agree that there are unacceptable behaviors that require outside monitoring and rectifying—because they have been ineffectually monitored and rectified from within—then what behaviors are they?

Conservative elements within the scientific community, grudgingly accepting the development of codes of conduct and systems of redress and accountability, would like to keep them to a minimum. For these (senior, powerful, self-interested) scientists, scientific misconduct should be restricted to three things, abbreviated FFP: fabrication, falsification, and plagiarism—that is to say, altering data unjustifiably, making things up, and stealing other people's words.

But what about other kinds of bad behavior on the part of a scientist, in a scientific context? What about torturing animals? What about hiring your unqualified girlfriend? What about shooting down someone's grant proposal or manuscript because you want to scoop them? What about stealing other people's ideas? What about retaliating against a whistle-blower? What about compelling a student to risk their health? What about lying to participants in a scientific study? What about hitting on your study subjects? What about doing research that can have no conceivable uses other than harming or killing people? What about listing someone as a coauthor who really didn't do much, if anything, on the paper?

The hard-liners maintain that these should be grouped separately from FFP: these acts suck, but they do not constitute scientific misconduct per se, they argue. They are some kind of "other" misconduct. They believe that the interests of science are best served by having as little regulation or intervention as possible—since some regulation or intervention is obviously necessary and unavoidable. The problem, clearly, is that they sound almost like they want to be able to *continue* sexually harassing, or lying, or retaliating in a scientific context, with impunity. Why else would they oppose classifying these as scientific misconduct and work to have them listed separately as rather more minor offenses? Is, say, torturing animals really not quite as bad as falsifying data?

In 1989, the National Academy of Sciences published "On Being a Scientist," a booklet intended for young scientists embarking on careers, in the wake of the recent fraud scandals and the clamor for reform. Most of the booklet dealt with subjectivity and error; it got cursorily to fraud near the end, to assure the reader that fraud is rare and you can't get away with it. Within a few years, the situation had changed so dramatically that in 1995 the academy was obliged to release a new edition, completely rewritten. Now it became a hard line for FFP, dismissing the suggestion that scientific misconduct might be expanded beyond FFP to include "other serious deviations from accepted research practices," as some institutions were now maintaining. The booklet did note that "this area of science policy is still evolving"—while striving as far as possible to maintain the equivalent of a creationist position, to keep things as close as possible to the way they always have been. And when it came to things like sexual harassment, "misuse of funds, gross negligence," and the like, those were somebody else's problems "and should be dealt with using the same procedures that would be applied to anyone."[37]

THE BALTIMORE CASE

If the Summerlin case heralded the modern era of scientific fraud, the Baltimore case was the beginning of the postmodern era—where wealth and power buy justice, the police become the criminals, a judgment of gross incompetence is a welcome relief to the accused, and a reign of confusion is the scientist's strongest ally, if not stock-in-trade.

The case centered on a paper by David Weaver et al., published in the high-profile journal *Cell* in 1986. Weaver himself has never been accused of anything awful and probably has long regretted being first author of the paper. The controversy centered on authors number 5 and 6. Number 5 was David Baltimore, Nobel laureate and perhaps the leading scientist and science administrator in the United States. Since the scandal broke twenty years ago, he has been president of Rockefeller University and Cal Tech. Number 6 was Thereza Imanishi-Kari, head of an immunology lab at MIT, then Tufts, accused of falsifying data.[38]

The whistle-blower was a post-doc in Imanishi-Kari's lab, Margot O'Toole. Struggling to build on Imanishi-Kari's contribution to the collaborative paper, O'Toole found laboratory notes that indicated her boss had actually gotten results with mice (those telltale mice again!) that were the opposite of what she had reported. That, of course, suggested a simple explanation of why O'Toole herself wasn't getting those results.

Following the procedures that many science bureaucracies had come to adopt in the previous decade, O'Toole detailed her discovery to the higher-ups. Choosing not to use the powerful eff word at first, she outlined what she found and what she thought it meant. She had little difficulty in convincing others that the Weaver et al. paper made numerous claims emanating from Imanishi-Kari's work which either were not substantiated or were wrong, but she could not convince them that this merited a correction or retraction. David Baltimore himself explained to her that the self-correcting process of science would take care of it, apparently missing the irony that he was now impeding that very process. "After trying and failing to deal with the matter unofficially, I went through official channels at M.I.T.," O'Toole later wrote.[39] She was told either to lay a claim of fraud or to shut up. Without a formal accusation of fraud, administrators breathed a sigh of relief and swept it aside.

O'Toole had reached the point where her idealism played to a stalemate against the realities of the scientific hierarchy, and she let the matter drop. Imanishi-Kari got a faculty research job at Tufts; O'Toole got a job at her brother's moving company.

But a former graduate student named Charles Maplethorpe had harbored similar suspicions about Imanishi-Kari's work en route to his doctorate, had taken them to senior administrators, and had been impressed by their lack of enthusiasm in pursuing the matter, so he had also let the matter drop. When he heard that O'Toole was on the same road, he put Walter Stewart and Ned Feder in touch with her. Stewart and Feder were scientists at NIH who had begun to take a decidedly unhealthy interest in the integrity of the scientific literature and in scientific fraud, a few years earlier. Stewart and Feder looked at the data O'Toole had found and became convinced that the conclusions were false and misleading.

Spurred on by the NIH fraud busters, the case got the attention of Congressman John Dingell's Subcommittee on Oversight and Investigations and became the subject of a congressional hearing on fraud at NIH. In January 1988, NIH appointed a committee of three immunologists to investigate the matter, but when it came to light that one had been a post-doc with Baltimore and another had coauthored a textbook with him, the committee was hastily disbanded and reconstituted. *Science*'s first write-up of the episode acknowledged that "the issue is extremely complex, and arcane even for immunologists." The scientists O'Toole had initially contacted about the work quickly wrote to *Science* that "there was (i) no sign of fraud; (ii) no evidence of misrepresentation; and (iii) no error that undermined the article's basic conclusion." That was the last time anyone would be so positive about the paper.[40]

A new NIH panel released its report at the beginning of 1989, finding significant errors in the Weaver et al. paper but no fraud or misconduct. Dingell called another round of hearings. NIH called another round of investigation. *Science* preemptively began making excuses: "Imanishi-Kari, whose native tongue is Portuguese, is notoriously difficult to understand." It quickly emerged that "there was plenty to support an argument that the preparation of the *Cell* paper was sloppy, and even flawed." Imanishi-Kari acknowledged not keeping particularly good records. Baltimore, who had adamantly been resisting a retraction of the article, reluctantly published two "corrections" to it in *Cell*, asserting that these in no way affected the conclusions.[41]

David Baltimore now marshaled his considerable scientific forces and went on the offensive, not only standing by Imanishi-Kari, but decrying the proceedings as a witch hunt. Stephen Jay Gould, likening Imanishi-Kari to Galileo, defended her by noting that "the paper contains some errors, and some evidence of poor record keeping. The more public charge of fraud cannot be sustained."[42]

But it was indeed sustained when the second NIH panel released its report in the spring of 1991. It found things to have been "falsified" and "fabricated" and that Imanishi-Kari in particular had said things to the investigators that "were known by her to be false, or were provided with reckless disregard for the truth."[43] Baltimore issued an apology and retracted the article in question, although Imanishi-Kari herself and another coauthor did not sign the retraction. Baltimore nevertheless maintained the integrity and correctness of the paper, even following the retraction. A distinguished biochemist called Imanishi-Kari's work "so sloppy as to insult the scientific method" and criticized Baltimore himself on four counts:

> He (1) failed to examine critically the quality and sufficiency of the data before publication; (2) failed to examine the data and report the possibility of error after serious criticisms were made; (3) instead organized an attack on his critics and discouraged publication of their views; and (4) did not subject the conclusions to further tests or check the reproducibility of what had been reported in a timely manner.[44]

Nature now editorialized about Baltimore's "admirable, but quixotic, loyalty . . . [which] would have been defensible if it had been based on a thorough examination of the data." Worse criticisms were to come, notably of Baltimore's "casuistical line of defence that the conclusions of the disputed article would and should stand or fall by others' confirmation of them, his frequent assertion that supporting evidence was, in any case, accumulating and his temporarily successful attempt to mobilize the scientific community in resistance to a supposed attack by the US Congress on freedom of science."[45] Baltimore resigned the presidency of Rockefeller University, licking his wounds.

Only six months later, federal investigators decided that prosecuting a criminal case against Imanishi-Kari would be a waste, because a jury would never be able to make sense of the skein of immunological, biochemical, and genetic data, record keeping, and claims and counterclaims that it would require to find her guilty. Baltimore promptly unretracted the retraction.

Two years later, in 1994, the Office of Research Integrity, which had replaced the Office of Scientific Integrity in the federal scientific bureaucracy, released its report and found Imanishi-Kari guilty on nineteen counts of scientific misconduct. Imanishi-Kari appealed the decision, and two more years later the U.S. Department of Health and Human Services overturned the ORI decision, clearing her of misconduct while nevertheless finding the Weaver et al. paper once again "rife with errors." David Baltimore assumed the presidency of Cal Tech, and Thereza Imanishi-Kari went back to work at Tufts.[46]

So what do we learn from the Baltimore case that we did not already know from Piltdown, Kammerer, and Summerlin?

First, a life lesson: with virtually unlimited resources and some public support, you can get away with almost anything. Like the O.J. Simpson case, which overlapped the tail end of the Baltimore case, any misstep by the prosecutors can be brandished by the defense—if the defense has the perseverance and the wherewithal and can keep the relevant public opinion confused and ambivalent. O.J.'s ill-fitting glove ("If it doesn't fit, you must acquit," said his lawyer, Johnnie Cochran) was paralleled by the Secret Service's analysis of Imanishi-Kari's data books, which also was unfortunately a bit sloppy. And the widespread feeling that justice is meted out differently to blacks and whites in the United States, and that consequently any case of a black man accused of murdering a white woman would never be fairly tried, had a parallel in the apprehensions of scientists about widespread "anti-science" attitudes in government, which David Baltimore took every opportunity to cultivate and exploit.

Second, a social lesson: it is not in anyone's interests to find fraud, and they will go to odd lengths to avoid it. The first people Margot O'Toole consulted told her after a cursory look that the work was fine, but it seemed to get worse and worse as time passed. Mostly they wanted to know if she was going to lay a formal charge of fraud, because that would get the engines of bureaucracy whirring. The work itself seemed to pass between fraudulent and not fraudulent with stunning ease, and, because of its highly esoteric nature and the difficulty in establishing the intent to deceive, the prosecution simply gave up in the face of adamant denials of fraud.

Third, a scientific lesson: incompetence is not a defense, and the end does not justify the means. For all the support that Baltimore was able to rally in the scientific community with his talk of witch hunts, he stumbled badly on his insistence that his colleague merely made mistakes and that other research rendered moot the issue of the competence and honesty of the Weaver et al. paper. After all, once you have established that your colleague's work is not reliable, it really doesn't matter why. If some scientists don't do good research, it is difficult to maintain that they should nevertheless still be employed and receiving grants, much less that you want to continue collaborating with them!

The problem with the "incompetence defense," then, is that it implicitly raises a question about the rest of their work and about your own judgment in standing by incompetent work. To say someone is a sloppy researcher whose work is riddled with mistakes is *not* a compliment, and it immediately raises the questions of why you are associated with such a person, how competent the rest of their research has been, and why they should remain at work. I can think of no other profession in which that would be tolerated.

Further, other people's research and conclusions are irrelevant. As the biochemist Paul Doty wrote in evident frustration, "The scientific literature would become irredeemably corrupted if this became accepted practice. The essential standard is that the evidence presented in a scientific paper is the bedrock on which interpretations and conclusions are built. If this connection is violated so that speculations drawn depend on subsequent investigations to prove them right or wrong, then the reporting of research would be reduced to a lottery."[47] In other words, as noted in chapter 6, the only relevant criteria for establishing the quality of scientific work is whether it is competent and honest; all else is polemics.

Finally, one's rank in the hierarchy makes a big difference in what one can get away with. William Bateson was a very senior figure, a former Cambridge professor and director of a major research institute—and as nemesis of Paul Kammerer, he was looking down on Kammerer. The mousekeeper who ended Summerlin's career never confronted Summerlin but sent the information up the Sloan-Kettering hierarchy; action was taken by Summerlin's superiors. Margot O'Toole, on the other hand, was blowing the whistle upward at her boss and was consequently always vulnerable to the ad hominem charge that she was merely a disgruntled employee. There was no "Science 911" phone number to call. Since science is a hierarchical bureaucracy, one is simply more likely to be a successful whistle-blower or nemesis from the top than from the bottom—because a low-ranking whistle-blower risks more, and a bureaucracy tends to coalesce around its higher-ranking members. And if the low-ranking nemesis or whistle-blower is not satisfied on the first go-round, the choices are few: become obsessed with it (as Helena Wachslicht-Rodbard did), or walk away (as Margot O'Toole tried unsuccessfully to do). That is presumably why a recent survey finds that scientists are generally reluctant to report a case of fraud, even when they are pretty sure it has taken place.[48]

Perhaps the ultimate case of high rank allowing a scholar to behave bizarrely without being called on it is that of Sir Cyril Burt, the first psychologist to be knighted for his work. Burt was a pioneer of statistical psychology and had a large impact on the development of the British public school system. Like many in the first part of the twentieth century, he believed strongly in the genetic basis of intelligence and in the IQ test as a way of measuring it. Unlike other psychologists, however, he was in a position to prove it, through the meticulous collection of the IQs of identical twins reared separately; their environments differed, but their genes did not. And he found their IQs to be very highly correlated with one another.

While Burt was alive, "there were certainly grave doubts although nobody dared to put them into print, because Burt was enormously powerful. . . . He would write a 50-page paper denouncing any criticisms." Shortly after Burt's death in 1971, however, American psychologists on both ends of the political spectrum had begun to identify oddities in his work, casting doubt on its scientific value. It soon came out that he had invented many of the twins, had invented several of his coauthors, had published pseudonymous articles praising himself and denouncing his critics in his own journal—and was, in short, as mad as a scientific hatter.[49] In desperation, other like-minded believers in the innateness of intelligence have subsequently tried to rehabilitate him, arguing that he was merely eccentric and careless—as if, once again, that made his work acceptable. Not only does it *not* render his work acceptable, but it goes against all that everyone agrees on concerning Burt—that he was a premier statistician and a fastidious worker! Certainly anyone capable of fabricating research assistants and praising his own work in their name is easily capable of fabricating data. And he seems to have done that, in spades—and gotten away with it.

MORAL CONDUCT AND SCIENTIFIC ETHICS

Although it is tempting to associate scientists' personal conduct with their professional conduct, the two are not necessarily connected. D. Carleton Gajdusek won the Nobel Prize in 1976 for his discovery of a new class of infectious agents, what we now call prions, the most famous example of which is the one that causes mad cow disease. In 1992 he wrote an excellent essay on the subject of the responsibility of scientists, especially geneticists, not to overhype their work. In 1997 he pleaded guilty to child sexual abuse and served a year in jail. W. French Anderson is widely known as "the father of gene therapy." Of course, gene therapy does not really exist, and other people working the field have long accused Anderson of being a credit hogger and spotlight hound, if nevertheless a brilliant researcher and administrator. And he, too, had an eye for underage sexual partners—and probably wished for less of the spotlight when he too was convicted of pedophilia in 2007 and sentenced to fourteen years in jail.[50]

John Buettner-Janusch for all intents and purposes opened up the study of Malagasy lemurs to the field of biological anthropology. In the late 1950s he became the first physical anthropologist hired full-time by Yale (I inherited his laboratory space there many years later). When he moved to Duke, he founded the Duke University Primate Center, now a world-renowned facility. His 1966 book, *Origins of Man,* is probably the last great critical, synthetic textbook of physical anthropology—before the market began to dictate that textbooks must be glossy and vapid. Then he moved to New York University to chair the anthropology department, until being arrested in 1979 for allegedly using his laboratory to make quaaludes and LSD and laundering the proceeds through a corporation he called Simian Expansions Limited. B-J served two and a half years in Florida at a minimum security facility known as "Club Fed," overlapping some of the Watergate miscreants, and was paroled. The one time I met him was during this interval, when he hosted a party at the annual meetings of the American Association of Physical Anthropologists in 1985. On Valentine's Day 1987, he sent boxes of poisoned Godiva chocolates to people whom he felt had done him wrong at different points of his career, and he spent the rest of his life in jail for attempted murder.[51]

In 1991, while B-J was in jail the second time, I invited him to review a book for the *American Journal of Physical Anthropology.* It turned out to be his last publication, for he died the following year. I got some grief about it, too; some colleagues questioned my decision to have a convicted felon review a book. My feeling was that he may have lost his moral bearings, and perhaps his sanity, but I was interested only in his critical anthropological faculties, which still seemed to be acute. I told those colleagues, "When you write a book as good as *Origins of Man,* I'll have you review a book for me, too."

I stand by that decision. He was unimpeachable as a scholar; his problems were as a citizen.

AN OCCASIONAL CONNECTION BETWEEN ORDINARY AND SCIENTIFIC DISHONESTY

On July 17, 1974, the *New York Times* ran the following editorial:

> The case against Dr. Charles G. Sibley, distinguished director of Yale's Peabody Museum, rests on the simple proposition that scientists, like politicians, are not above the law. An outstanding ornithologist, Dr. Sibley has been fined $3,000 in a civil procedure for having systematically imported birds' eggs and egg whites that he was not licensed to import.
>
> Scientists affirm the importance of Dr. Sibley's work, which involves a new method of classifying bird species by the protein content of the albumen. If his offense had been no more than the occasional and unsolicited receipt of an egg taken without a permit, he would perhaps be justified in complaining of "persecution" for merely technical violation of the Lacey Act—a statute which contains, among other provisions, a ban on the importing of any animal or animal part taken contrary to a foreign country's wildlife protection laws.
>
> Unfortunately the case involves more than that. Dr. Sibley appears to have used the services, in England, of an organized ring of illegal operatives, whose raids included the taking of eggs from the nests of such rare birds as the peregrine falcon, the stone curlew and the ringed plover. Dr. Sibley, it was charged, willfully received some of the material as well as the eggs of birds less endangered but nevertheless not stipulated in permits issued to him.
>
> No doubt the temptation to circumvent bureaucratic red tape was strong, and Dr. Sibley's activities, unlike those of most violators of the Lacey Act, involved no personal profit. Nevertheless, as clear and deliberate evasions of the law, they cannot be justified by scientific purpose. The arrogance of science is no more appealing than the arrogance of commerce—or of government.

"The arrogance of . . . government" was a reference to the Watergate scandal, that infamous "third-rate burglary" of recent memory. The fine paid by the Yale professor was a plea bargain, the back end of the Fed's dropping criminal charges against him. And yet, although the story made the *Times*, oddly enough it didn't make the science journals. Sibley was indeed a respected ornithologist; more than that, he was a pioneer in the field of molecular evolution—examining the structures of molecules (proteins and DNA) to understand the relationships among species. But the scientific community was largely spared news of this scandal. It was discussed not in the pages of *Science* but in the pages of *Sports Illustrated,* of all places—where falconers were rather more indignant about it than were molecular evolutionists.

Indeed, the *Times* had erred on one critical point. It was *not* the case that Sibley's involvement in smuggling the eggs of endangered species out of Europe and into his electrophoretic apparatus "involved no personal profit." The profit involved a different currency than most illegal trafficking in rare animals does, but Sibley of course profited from these acts through the currency of publications, grants, and the professional stature they buy. Later that year, he was even made first vice president of the American Ornithologists' Union.

And so the scandal passed, and Charles Sibley remained on a pinnacle of science, a distinguished professor at a distinguished university, publishing distinguished research on molecular evolution—until he was caught egregiously falsifying data fifteen years later.

This was interesting for several reasons, not the least of which is that by now Sibley had been elected to the National Academy of Sciences.

Now, in 1987, as in the Lacey Act issue, some of Sibley's colleagues were rallying to defend him. In fact, the *New York Times* editorial back in 1974 had been stimulated by its own news story a few days earlier, rife with quotes from another prominent biologist minimizing the allegations—suggesting that the acts of stealing and destroying the eggs of endangered species were being overblown and that Sibley was actually being set up by "extreme conservationists." Now, some years later, once again his friends emerged to say that it was all a big mistake, that he was being set up and persecuted, and that even if there was a technical breach of conduct it wasn't that critical.

By this time, however—1988—Sibley had already admitted that his published data had been subjected to alterations, the fact and nature of which had been consistently withheld from reviewers and readers. In 1993, two of Sibley's friends could actually write in his defense that

Sibley and Ahlquist's methods "made very little difference to inferences from the complete data"—although Sibley himself had by then already admitted the opposite, that it was "virtually certain" that without those unreported data alterations their conclusions would actually have been quite different.[52]

What, I wonder, could motivate scientists to defend a colleague accused of data falsification by publishing such an easily demonstrable falsehood themselves? Obviously, they must have perceived a lot to be at stake.

Sibley was a pioneer in the application of a technique called DNA hybridization to the question of the relationships among species. In the mid-1980s, under his leadership, a handful of other scientists had adopted the technique and were favorably reviewing each other's grants and papers. Outside the circle of friends, however, there was some very justified skepticism. Data were exceedingly hard to come by if you were outside the circle, because, despite vigorous and uncritical promotion in the secondary literature, the technique worked well only for a fairly restricted range of phylogenetic questions, or if one took very subjective liberties with the data. The technique was actually often quite crude and most frequently yielded interesting but unconvincing results. The trick was—if you wanted to apply it more broadly—to transform the ambiguous results into convincing results and to conceal that transformation from potential skeptics, like external reviewers and readers.

A 1991 review in *Science* of the magnum opus *Phylogeny and Classification of Birds,* by Sibley and his protégé Jon Ahlquist, explicitly complained,

> Sibley and Ahlquist have modified an unspecified number of their . . . values, and the effect of these changes is unknown. Although the authors discuss the principles behind them, the alterations are *a posteriori* and subjective. The reader cannot decipher how "corrections" affect a given data set or conclusion. This is a nagging problem because there are actually two overlapping kinds of data manipulation to worry about.[53]

So Sibley and his protégé published numerous papers through the 1980s that consistently omitted crucial aspects of their data analysis;

they were very reluctant to release their data to potential critics; those unreported analyses were discovered by others; they affected the results significantly but had been withheld from the reviewers; and they were "subjective." By the mid-1990s, the circle's worst fears had been realized, and the plug had been pulled on funding for this line of research, on the basis of the Sibley controversy.

But what exactly was controversial here? What kind of science makes crucial and subjective changes to the data, fails to include that information, and withholds its data from the scrutiny of others? Only one kind of science—fraudulent. Slam-dunk.

And yet nobody ever convicted Sibley of anything. In fact, nobody ever formally looked into it in order to convict him of anything. The problems with the work had been discussed in the primary literature (the *Journal of Human Evolution, Cladistics,* and the *American Journal of Physical Anthropology*) as well as in the pages of generalist journals *(Science, Nature, Scientific American,* and *American Scientist).*[54] Although the work had been done at Yale, one of the principal accusers was at Yale, one of the principal defenders was at Yale, and even the student science magazine at Yale had written it up, the university itself never formally looked into it. Why not? Well, Sibley and Ahlquist had already left Yale, so Yale couldn't very well discipline them. What purpose would an investigation possibly serve, aside from generating some very bad karma? The National Science Foundation didn't seem terribly interested in getting its money back, either.

Given that it was nobody's job, then in whose interests might it actually be to conduct an investigation and formally declare the existence of scientific fraud, which was unlikely to have much effect on the alleged perpetrator, who in any event was a distinguished senior scholar? Perhaps it was in the interests of the National Academy of Sciences, which had recently elected Charles Sibley to its elite ranks on the ostensibly honest merits of his work.

That was precisely the question I raised in the pages of the *American Scientist* in 1993.[55] (Yes, I had been one of Sibley's principal nemeses; and it was this episode that was largely responsible for getting me interested in the anthropology of science in the first place.) Sibley and Ahlquist blustered back in the next issue, "Dr. Marks urges the National Academy of Sciences to conduct an investigation into our alleged crimes against science. We shall suggest such an investigation to the National Academy of Sciences Home Secretary."[56] Just in case they later forgot about it, I made the same suggestion in a letter and package to that home secretary, the distinguished botanist and ecologist Peter Raven.

This is what the National Academy of Sciences home secretary wrote back to me, on October 25, 1993:

> Thank you very much indeed for your letter and the enclosures. I was extremely interested in what you had to say in reading the enclosures. It is obviously a very complex case and, as I am sure you understand, the National Academy of Sciences would not undertake to conduct a formal review of the activities of its members as a matter of general principle, lacking the judiciary machinery to do so properly. I would add, however, that no one is elected to the Academy for a single piece of work, and thus it is incorrect, as a matter of principle to say that "this is the work that ultimately resulted in Sibley's election to the National Academy of Sciences. . . ." In summary I was very interested in the material that you sent. We will be conducting no investigation.
>
> Yours sincerely,
> Peter H. Raven

I have nothing but admiration for this great scientist, but ultimately I turned to primatology to make sense of his letter, and recalled Mizaru, the first monkey, the one with his hands over his eyes. Then I moved on.

EIGHT The Rise and Fall of Colonial Science

As the American nation expanded westward in the early nineteenth century, expelling, impoverishing, and exterminating the former inhabitants of the land, it soon came upon a remarkable series of geological phenomena. Near St. Louis, for example, sits a group of enormous conical mounds full of skeletal and cultural remains—a site now known as Cahokia. They seemed almost to be the overgrown remains of ancient towns, and clearly the products of human creativity.

But the activity of what humans? Surely not the filthy savages we were busily displacing, dispossessing, and decimating. If their ancestors had made such large-scale and impressive things, it would seem as though they had a rather noble pedigree, and their suffering at our hands might seem almost . . . well, criminal.

No, it must have been someone else who constructed those mounds. Somebody different—in fact, perhaps somebody who actually had been done in by the Indians we were killing. So maybe it even kind of served them right. But who were the mysterious ancient wonderful people who actually built the mounds? Hebrews? Vikings? Druids?

Thus was American archaeology born, from the womb of an idiotic, racist question that would not be finally settled until the end of the century. Of course, it was the ancestors of the Indians themselves, and nobody else, who had built the mounds. The problem, such as it was, lay simply in the reluctance to acknowledge that the indigenous people, who were being aggressively dehumanized in order to rationalize what was being done to them, could possibly have done something so impressive.[1]

Not coincidentally, a similar myth developed in southern Africa late in the nineteenth century, around the ruins known as Great Zimbabwe. There, a local despot of almost unimaginable wealth and power, Cecil John Rhodes, was always up for a new way to rationalize his plundering of the indigenes. With the Oxford scholarships and two countries (Northern and Southern Rhodesia, now Zambia and Zimbabwe) named after him, the magnitude of his clout is readily attested—with the aid of a fortune from diamond mining.

After a German explorer called renewed attention to the site, with great stone walls and buildings, Rhodes financed the first archaeological study of Great Zimbabwe, by J. Theodore Bent. Bent obliged by interpreting Great Zimbabwe as the product of voyagers from the Near East, with a Biblical nod to the Queen of Sheba; he argued that the architecture of Great Zimbabwe was "not in any way connected with any known African race." As a premodern American anthropologist put it, "No negro or negroid race ever built stone walls voluntarily."[2] This view was quickly disseminated in the immensely popular novels of H. Rider Haggard, who specialized in exotic fantasy adventures (his most famous novel featured a beautiful white queen in Africa, known to her subjects as "She Who Must Be Obeyed"). In several other novels—notably *King Solomon's Mines* (1885) and *Elissa* (1899)—Haggard reproduced the idea that a great lost non-African race was responsible for the admirable works of the Dark Continent. It was a subtle but crucially political point: Cecil Rhodes might simply be reclaiming the wealth of the Bible, and its ancient colonies, for the races who lived thousands of miles to the north.

Archaeologists at the turn of the century were consequently a bit more divided than the public, whose knowledge of southern African prehistory was largely filtered through Rider Haggard's racist assumptions. By the time that David Randall-MacIver published his archaeological research in 1906, it was clear that there had been extensive trade down the east coast of Africa, but that the trade goods were about twenty-five hundred years younger than King Solomon. Moreover, the form of the ruins themselves suggested stone versions of the local African villages, and the artifacts themselves looked quite familiar to anyone knowledgeable about southern African material culture. Nevertheless, some still protested angrily that Great Zimbabwe was "altogether beyond the capacities of existing Kafir races." Further research by Gertrude Caton-Thompson in 1929 established that it was both indigenous and medieval: "No object . . . bears, in my opinion, the impress of either remote antiquity or of foreign occupation."[3]

And yet it remained in the government's interest to maintain the fiction that there was ambiguity and mystery surrounding the origins of the site and its artifacts.[4]

History is political, and origins are histories. Consequently, it should be neither surprising nor scandalous that, when science and history converge, they are invariably vested with assumptions, interpretations, and manipulations that give them political weight. This is true even for something as apparently wackily innocuous as Erich von Daniken's 1970 best seller *Chariots of the Gods?* which spawned a large number of derivative paperbacks with similar cover art, typefaces, and ideas.

Could extraterrestrials have built the Great Pyramid at Giza? Sure, they could have; but the chances that they actually *did* are about the same as their having built the Eiffel Tower. After all, the ancient Egyptians not only seemed to think that they did it themselves, but also left us their learning curve—in the form of earlier tombs, such as the Step Pyramid and the Bent Pyramid. But more relevant is the assumption upon which the question is founded—that there is some kind of problem with the idea that ancient Egyptians are the ones who actually did build the pyramids. In fact, it is the same problem that underlay the questions "Who built the mounds?" and "Who built Great Zimbabwe?"—namely, that there is some reason to doubt that the indigenous peoples actually could have done it themselves. In other words, the question is framed in such a way as to presume that the indigenous peoples were constitutionally incapable of doing it themselves, and therefore someone else either must have done it for them or helped them out crucially. Otherwise, why even bother asking the question?

What runs through all of these rhetorical questions is the presumption of some deficiency in the nature of the people themselves. Whether that undergirds the work of pseudoscientists like Erich von Daniken or of scientists like the geneticist Bruce Lahn (chapter 6), it is fundamentally a racist supposition, and as such it requires a lot more justification than do other scientific assumptions. After all, it is more likely that the intellectual deficiency lies in the head of the questioner rather than in the heads of the indigenous peoples—it's not that they were stupid and couldn't have done it; it's that they were cleverer than you and did it. Indeed, in most such examples, the question of how they did it lies not so much with the ideas themselves as with the manpower and motivation for doing so. In other words, it is not so much about intelligence as it is about labor and logistics—getting lots of people to want to do it and providing the social infrastructure that allowed them to.

"How did they do it?" thus becomes an answerable question of social process rather than an unanswerable question of mysterious unknown races and lost gifts from beyond. In other words, it becomes science.

PSEUDOSCIENCE AND HISTORY

Generally credited with being the first to attempt an explanation for human history in genetic terms was a nineteenth-century French nobleman, Count Arthur de Gobineau. His major work was called *The Inequality of Races* (1854), and for this reason he is often known as the father of scientific racism. Obviously, that is not intended as a compliment. Gobineau was attempting to mount a last-ditch defense of the declining ancient European hereditary aristocracy by producing a unified theory of race and civilization in which the aristocrats played a central role.

His theory was that civilizations rise as a function of the intellects of their individual members or, more specifically, of their leaders, and that civilizations fall as that elite blood is dissipated through interbreeding with the masses. Gobineau argued that the white intellect is higher than the black or yellow, and that within the white race the "Aryans" are the intellectually superior subrace. Moreover, of his ten identifiable civilizations, Gobineau attributed at least seven to Aryan blood and identified no civilization at all in sub-Saharan Africa.

Gobineau's thesis was tightly argued. Past civilizations, and by implication the fate of the present one as well, were governed by the purity of blood of the aristocracy, whose position in the social order was ordained by nature. Social change and mobility, as well as social equality, were contrary to nature. The future of civilization lay in the recognition of the unequal abilities of races and in the preservation of the social hierarchy from which it arose. Gobineau's argument would be reiterated and adapted in the writings of Houston Stewart Chamberlain in Europe (*Foundations of the Nineteenth Century,* 1899) and Madison Grant in the United States (*The Passing of the Great Race,* 1916).

In his own lifetime, Gobineau's racial theory of civilization was eclipsed by the arrival of Darwinism, which seemed to imply a much more unstable biological nature than Gobineau supposed. But that very biological instability could be inverted; indeed, the saddest aspect of the emergence of modern genetics is the renewed vigor it began to give to racist pseudo-Darwinian explanations of history. The geneticist Charles Davenport could articulate the qualities of the German gene pool in 1911, before the two world wars: "Germans are, as a rule, thrifty, intelligent, and honest. They have a love of art and music, including that of song birds, and they have formed one of the most desirable classes of our immigrants." In concert with the age, Davenport had far less flattering things to say about Jewish immigrants to America, even if they also liked songbirds.[5]

Interestingly, the gene pool of Ashkenazi Jews, which had been vilified by American geneticists and psychologists in the 1920s and by their German counterparts in the 1930s and 1940s, made a remarkable comeback after the Six-Day War in 1967. By 1969, C. P. Snow (of "The Two Cultures") was singing its praises: "Is there something in the Jewish gene-pool which produces talent on quite a different scale from, say, the Anglo-Saxon gene-pool. I am prepared to believe that may be so."[6]

Later that year, the English botanical geneticist C. D. Darlington published a synthesis of human history in terms of inbreeding, outbreeding, gene flow, and the imaginary endowments of local gene pools. As a geneticist, he probably should have been more attuned to the inherent difficulties in trying to infer properties of the genotype from observations of phenotypes, but when politics and history are at stake, as noted above, all bets are off. Many were impressed by his erudition, but most were appalled by his crudely disguised bigotry. As the immunologist Sir Peter Medawar put it, "It has always seemed to me that strong adherence or repugnance to the dogma of genetic determinism raises a psychological rather than a scientific problem. This is a matter which deserves more attention than anyone has given it."[7]

But, like the phoenix of legend, even in the twenty-first century biologized theories of human history can rise from the ashes. A journalist for the *New York Times* resurrected the gene pool as the engine of social history in a popular book in 2006, identifying a presumptive genetic basis for Jewish intelligence, Chinese ping-pong, and the "cultural stagnation" of non-European peoples. While claiming to popularize anthropological genetics, the book was nevertheless appropriately trashed by leading anthropological geneticists.[8]

The trouble with genetic theories of human history is that they are theories of fate, not of history. They look at what is and reconstruct why it had to be so, without recognizing the precariousness of history and appreciating that it could very easily have been otherwise. A genetic theory of history is circular: building a gene pool of the past from the accomplishments of the present, it argues that the present is simply a consequence of the inferred genetic endowment. And the circular theory produces a teleological historical model, for history itself becomes simply an unfolding or expression of those imaginary endowments.

In this sense, then, genetic theories of history are largely valueless. This is not to say that genetics and history have never intersected. It may well be the case that, for example, Europeans were unable to overrun sub-Saharan Africa demographically, as they did America and Australia, because of their greater sensitivity to endemic diseases such as malaria and yellow fever, with which African gene pools had coevolved.[9] It is also conceivable that the survivors of the Black Death in the fourteenth century, which may have killed upward of a quarter of the people of Europe, represented a nonrandom sample of the gene pool—those with some fortuitous genetic resistance. These examples, however, constitute very different invocations of genetics and its possible relationship to human history.

To understand the social facts with which history presents us, then, we need to invoke social processes, not biological ones. This late nineteenth-century recognition is as fundamental to social science as uniformitarianism is to geology or the cell theory to genetics. To the extent that there may be exceptions (such as mass extinctions caused by extraterrestrial impacts, or viruses as units of life), they are rare and interesting principally as second-order generalizations.

That is why another recent book, Jared Diamond's *Guns, Germs, and Steel,* in trying to be politically correct, commits the same error as its antagonists. It begins by telling us that the indigenous peoples of "stone age" New Guinea are not innately dumber than Europeans, but smarter! Why? Because avoiding the leading causes of death in tribal society—murder, mayhem, accidents and starvation—actually selected *their* imaginary intelligence alleles. That assumes, though, that intelligence is a fairly unitary phenomenon, that the selective forces were different enough in different parts of the world and consistent enough for long enough to make a difference, and that other contributions to the probability of survival don't overwhelm whatever intellectual differences may exist within each population. In short, a pretty tall order.

Indeed, all of these assumptions are probably false—as was appreciated in the backlash against the eugenics movement in the late 1920s. Even nonscientists like H.L. Mencken recognized that prominence—however defined—is highly situation dependent, and that the times, so to speak, make the man. There is no reason to think that George Patton would have enjoyed the same success in Gaul that Julius Caesar had in the first century B.C. with the Roman legions, nor that Caesar would have been able to conquer the same territory in the 1940s at the head of the American Third Army. There is no reason to think Marlon Brando would have been a great actor in the eighteenth century, or that David Garrick would have achieved prominence in the field in the twentieth. That talent will shine through, transcending time and space, is nothing more than a conceit.

Further, there is no reason to think that Isaac Newton would have developed the theory of relativity if he had lived in the twentieth century, or that Albert Einstein would have developed classical mechanics had he lived in the seventeenth. William Harvey would not have discovered the germ theory had he lived in the nineteenth century, and Louis Pasteur would not have discovered the circulation of the blood in the seventeenth century. Sooner or later those things would have been discovered, for the purpose of science is the production of knowledge, but the discoverers would have simply been other smart people drawn from the same ranks.

The problem here lies in the cultural assumption that prominence, or gift, or genius is somehow the limiting factor in cultural histories. But it isn't. People's gifts expand to fill the needs of the time and place, and of course the crucial aspect of culture as a human adaptation is that it gives groups of people the ability to achieve as a collectivity what they cannot achieve as independent organisms. No one, after all, can build a computer from raw materials, however vast their genius may extend.

In short, the fallacy lies in invoking pseudo-Darwinian explanations to rank human groups intellectually; it doesn't matter which one is placed in the top tier.

Guns, Germs, and Steel then goes on to eschew racist explanations for European political dominance—identifying the flaws of other peoples not in their minds or bodies but rather in where they live. Instead of adopting a biological determinism, then, the book presents a geographic or environmental determinism. Once again, whether the causes are endogenous or exogenous, somehow history is being replaced by fate in this model. The facts of human history are as they must be; they could have been different only if the world itself were different.

Historians are reluctant to embrace such a thesis, since it is, after all, largely anti-intellectual—dismissing what the real experts have to say about how human history works.

Cultural change actually has three critical components, in addition to the crude limitations imposed by the environment, the randomness of events, and the occasional ability of a single individual to control and focus social resources with significant and unforeseeable results. The first of these components of cultural change is the general unwillingness of people to abandon what they have been doing, and what has traditionally worked for them, when there is no obvious advantage in changing (e.g., inches and feet, ounces and pounds in the contemporary United States). In fact, most aspects of culture are remarkably conservative. Considerably more time elapsed between the cave art of Chauvet and the nearly identical cave art of Lascaux, for example, than between the cave art of Lascaux and the pop art of Warhol! In other words, people generally have to be highly motivated to alter what they feel comfortable with.

The second component is the principal exception to the first, namely, that technology is largely autocatalytic. A species that has evolved biologically to survive technologically will naturally be tending to play with what it already has, to apply it to other problems, and to transform it in the process. If nothing comes of that activity, then so be it; but if it works, they keep it—and from a standpoint in the future, the activity will look like progress or improvement.

The third component is contingency—that the possible steps you can take are largely dictated by where you are stepping from. No amount of genius can make an iPod in a copper-smelting society. There is simply too much to be imagined, invented, developed, believed, produced, and institutionalized. The best you can do is apply what you already know to the production of other metals with different properties—like, say, tin or iron. But, of course, that is much easier said than done, and you would probably need to be very highly motivated to waste your time on it.

Taking history as a consequence of nature, either genetic or geographic, is to situate it within the realm of the inevitable—as if this was the way it had to be, representing history as destiny—which is an exceedingly primitive approach to the subject. The primitiveness lies in taking European political hegemony as somehow destined rather than as contingent. Rather than asking what's wrong with other people's history, we need to ask how ours happened to turn out this way; for it was mandated neither by the genes nor by the environment but by the choices and acts of people, as individuals and as groups—which could easily have been different, and the effects of which, at any point in time, are likely to be transitory. That is what makes history interesting and different from science.[10]

US AND THEM

The precariousness of history brings Europeans and their descendants that much closer to the other peoples of the world. If the history of the world is caused by defects in their genes or their continents, then the political and economic domination of those presently in control was more or less inevitable. If history is governed by human agency rather than by genetic karma or geographic kismet, then "we" are that much more like "them." Indeed, under slightly different circumstances "we" might even have been "them."

Bringing "us" and "them" closer together is what anthropology has strived for in the century and a half or so of its existence as a profession, so it is not to be taken lightly. Indeed, even a good chunk of the history of science is bound up in just that distinction. The process of colonial expansion, both in the Americas and in the rest of the world, involved science and scientists. They were studying the people, plants, animals, and minerals of the world for the sake of knowledge—but they also were subsidized by governments and companies looking to exploit the diverse resources the scientists were discovering.

Some of those resources involved the land the people used to own, or the knowledge they possessed about it—the most classic forms of colonial exploitation. But other resources included the very bodies of their ancestors. As the great museums began to store up the artistic and scientific treasures of the world, some of those treasures included human remains. As George A. Dorsey bragged in the *American Anthropologist* in 1900, on behalf of the Field Museum in Chicago:

> Naturally much osteological material of great ethnic value was procured . . . , as well as with many of the collections obtained by purchase. As a result the department was in possession of skulls and skeletons from Alaska, the Northwest coast, and several of the Plains tribes; from

Ohio, New Jersey, and Arkansas mounds; from prehistoric graves in Costa Rica, Colombia, Peru, Bolivia, and Chile; and . . . many specimens from America, Europe, Asia, Africa, and the Pacific islands.

In the division of physical anthropology more than 150 skeletons were accessioned, the most important single collection being one of fifty-two Papuan skulls from Gazelle peninsula, New Britain, received in exchange from Dr Parkinson.[11]

One notices quickly the sterile vocabulary employed: *osteological material, specimens, procured, obtained, accessioned.* It may not sound quite as scientific, but it sounds rather like the remains of dead bodies were being dug up and sold to the museum, perhaps even by someone named Igor. Or, in the care of "Dr Parkinson," they were traded like baseball cards. Dorsey doesn't say what Parkinson got in exchange for his Papuan skulls, but it does sound as if the Field Museum got the equivalent of two David Ortizes for a Derek Jeter.

The crucial point here is that the scientists implicitly regarded the Papuans' ancestors' remains as somehow less sacred than their own. You can't really imagine Professor Dorsey exhuming his grandparents in exchange for some Papuan skulls. Indeed, the macabre nature of that scenario was adopted in Tony Hillerman's 1989 best-selling novel *Talking God*—as an Indian activist sends a museum spokesperson her grandparents' bones, to emphasize their equivalence.

The phrase *colonial science* has been used in several different ways by historians, often referring to science done in a particular era, or in a particular region. The historian Susan Lindee, however, uses the term in a broader sense, pertaining to the social and political relations that exist between scientists (as the agents of a dominant and foreign power) and the local people. In this sense, she is able to explore the studies by American geneticists upon post-Hiroshima Japanese, although both the United States and Japan were arguably themselves colonial empires.[12] It is in this cultural sense—as the interaction between two different groups of people in a hierarchical power relationship, with the scientists on top—that I use the phrase.

With *colonial science,* then, I am referring to science carried out with a particular mindset: namely, that some people—because of their political, economic, or social standing, as a result of where they come from—are there to be taken advantage of and not necessarily to be treated as fully equal beings. They are other kinds of people, and scientists therefore may feel fewer responsibilities to those kinds of people than to their own. In other words, the scientist and the subject are recapitulating or embodying the relationships between their nations or peoples.

Colonial empires sent out their scientists to assess the local flora and fauna, as well as minerals, in order to ascertain what might be of value in the new lands.[13] They also used science to create products for export. As Cecil Rhodes put it back in 1890, "We must find new lands from which we can easily obtain raw materials and at the same time exploit the cheap slave labor that is available from the natives of the colonies. The colonies also provide a dumping ground for the surplus goods produced in our factories."[14] These days, it tends to be biotechnology being exported to the third world, often addicting farmers to new genetically engineered seeds but without materially increasing their living standard, or at best creating ambiguous new relationships between farmers and their land.[15]

Scientists especially adopted a colonial approach to the people themselves, in order to study them, shall we say, "scientifically." The "four-field approach" in American anthropology began in the late nineteenth century as a way of comprehensively othering the now-pacified Indians and dispassionately examining their ways of life, languages, material remains, and bodies—that is to say, their cultural, linguistic, archaeological, and physical natures. Likewise, European social anthropology originated as a handmaiden of the ruling powers, especially in Africa, even if the second generation of British ethnographers rebelled and often bit the colonial hand that was feeding them.[16]

COLONIAL SCIENCE

In general, colonial science is characterized by one or another of a suite of features. The most infamous is the dehumanization of other peoples (usually indigenes, in the present context)—that is to say, treating them as lesser beings. In 1904, for example, while George M. Cohan was on Broadway singing "I'm a Yankee Doodle Dandy" in his show *Little Johnnie Jones*, an African pygmy called Ota Benga was in St. Louis at the World's Fair. He was visiting, but not as a tourist; rather, he was a display, brought there by an explorer and entrepreneur named Samuel Verner. At the end of the fair, Verner took him back to Africa, and thence to New York, where he arranged to have Ota Benga displayed in the Bronx Zoo, along with an orangutan, in the primate house. In fact, integral to the negotiation was the secretary of the New York Zoological Society, our old friend Madison Grant.

There Ota Benga stayed for most of the month of September 1906, while African American ministers lobbied with the zoo and the mayor's office to have him removed from display. Finally he was transferred to an orphanage and removed from public display. Ota Benga lived his final years in Virginia, where he killed himself in 1916.[17]

As sad as that story is, the issue of dehumanizing people underwent a bizarre renaissance in the 1990s, with the humanization of apes by activists in the animal rights movement. In their zeal to make the apes seem more human, and thereby deserving of rights that transcend "animal" rights, several of these activists and philosophers casually began to associate the apes with autistic or mentally handicapped children. But such a comparison is false (or at least true only in a very narrow and arbitrary sense—performance on certain cognitive tests) and irrelevant (in modern society, rights are not predicated upon one's cognitive level; both geniuses and morons are free to vote, drive, and shop). Worse yet, the comparison recapitulates the very criticism voiced against the captors of Ota Benga: they regarded a human as if he were an ape and by implication denied him fully human status. Obscuring the full humanity of the weakest members of society is hardly a valid argument for raising public consciousness about apes.[18]

Another theme of colonial science involves the removal of goods and resources, without appropriate compensation, from other lands and peoples who are their presumptive owners. This behavior has a long and diverse history. Ancient empires would bring home loot from their conquered territories. Napoleon's army retrieved many objects of art during its Egyptian campaign and sent them back to Europe. One thing they left behind was the Rosetta Stone, which was captured and sent home by the English instead and has spent the past two hundred years in the British Museum—just a few hundred feet from the marble friezes brought back from the Parthenon in Greece by Lord Elgin.

The flip side of this form of colonial science is repatriation. In the United States, heightened sensibilities in the 1980s led to the passage of the Native American Graves Protection and Repatriation Act (NAGPRA) in 1990, which mandated the census and possible repatriation of biological and cultural objects that might have been improperly acquired from Native Americans. Although initially opposed for being "anti-science," the act came to be appreciated for the human rights legislation it was intended to be.

On the other hand, the compensation issue is one that has been more successfully opposed. In 1951, the first human cell line was successfully established in culture, derived by researchers at Johns Hopkins from the cervical tumor of a black woman named Henrietta Lacks, who died from her cancer at age thirty-one, never having been asked for permission, nor having consented, to have anything done with her cells. Probably 99 percent of what we know about human cell biology is derived from her cells—some of it inadvertently, since her cells have also overrun cultures derived from other lines. They're tough cells—they've been through the polio vaccine and they've been in outer space. Wherever they encounter other human cells they kick butt. Careers and fortunes have been made on HeLa cells, but neither Henrietta nor her descendants have ever seen a nickel from them.

Neither did John Moore, who developed a cancer of the spleen in the 1980s while working in Alaska and was cured by a specialist at UCLA. He didn't consent to anything either, but when his doctor insisted that he return to Los Angeles periodically and the doctor would pay for it, Moore became a little suspicious. He soon made a significant discovery, although not a scientific one. He learned that the doctor had patented a cell line derived from his cancer and made a pretty penny from it. With modern middle-class American sensibilities, Moore sued. In 1990, the California Supreme Court ruled against him. Even though without

Moore there would have been nothing of value, his only contribution was the raw material; the doctor had transformed it into something valuable. Giving Moore a financial interest in the products of his own body might "destroy the economic incentive to conduct important medical research"—although just how the progress of science itself would be retarded by letting the patient share in the profits was never made clear.[19]

A third aspect of colonial science is the manner in which it is self-perpetuated, by failing to train the people themselves, keeping them in the dark about crucial aspects of the science, or otherwise rendering their contributions invisible. Consider paleoanthropology, traditionally dominated by Europeans and Americans working in other parts of the world—although at the leisure of the local government and with the assistance of local laborers. Louis Leakey, who fancied himself a "white African," was working with teams of black Africans from the beginning (in the 1920s), but their names began to appear in his scientific publications only much later. Heselon Mukiri was his most trusted assistant and was well known to be Leakey's right-hand man, but it is not easy to find his work formally acknowledged in Leakey's copious primary scientific literature until a comment in a 1961 paper, thanking "my senior assistant, Heselon, on whom the bulk of work in this very long season has fallen."[20] A 1964 paper in *Nature* by Louis and Mary Leakey names as contributors to the endeavors "Miss M. Cropper," "my *[sic]* son, Richard Leakey," and "Mr. Glynn Isaac." It goes on, "Mrs. Isaac, Mr. Richard Rowe and Philip Leakey also took part, as well as a number of our African staff." One member, however, was singled out a bit further down, namely, "one of our African staff, Mr. Kamoya Kimeu"—who would later become Richard Leakey's ace fossil finder.[21] Along with Bernard Ngeneo, he would be regularly acknowledged for his contributions, beginning in the late 1960s.

In the postcolonial age, it is not uncommon for excavation permits for Europeans and Americans to be tied to the identification and training of students from the African or Asian country in which they want to work. This not only helps to staff the museums, universities, and ministries with competent scientists, but also creates a class of scholars who can work to balance the interests of science against the interests of the nation.

Once again, however, the practice of trying to keep people in the dark about science, so that they don't know the value of what you are taking from them and feel that they need you, occurs across class lines as well as cultural lines. In the case of Henrietta Lacks, not only was her family not told about how valuable mom's cells were, but the scientists created a fictitious persona named "Helen Lane" as the source of the HeLa cell line so they could plausibly deny any association with Henrietta Lacks herself. Decades later, it was discovered that HeLa cells had contaminated other human cell lines, were overrunning them, and could not be reliably distinguished from the other cell lines. So cell biologists devised a way to assay specifically for Henrietta's genotype—but they needed fresh cells from Henrietta's descendants to develop it. So they contacted the Lacks family and told them something about cancer and their mother and heredity. The family, desiring to help cure cancer and not get it themselves, readily gave samples. The cell lines were saved; reputations, livelihoods, and investments were restored. Don't even bother to ask what thanks the family got.[22]

In primatology, modern researchers are quite far ahead of the curve in educating and training local people. A major reason is that conservation is now the central issue in primatology—it doesn't help to know what monkeys do if there are no monkeys—and conservation is an issue for the local people as well as the monkeys. For conservation programs to be effective, there must be not only indigenous interest in the primates but an economic incentive for the people to participate in the program. Telling Africans not to poach gorillas is one thing, but presenting them with other career options—along with the message not to poach gorillas—is quite another. That recognition represents the largest difference between primatology as practiced by Dian Fossey and as practiced by her successors—after Fossey's grisly death, most likely at the hands of the African people to whom she was trying to deny a livelihood poaching gorillas.

One of the most successful postcolonial endeavors in primatology is the work of Karen Strier, who has been studying the woolly spider monkey, or muriqui, in Brazil since 1982. While her own professional fortunes may reasonably be said to lie with training American students and repopulating primatology programs with her intellectual descendants, the fortunes of the muriqui lie in the hands of the Brazilian people. Consequently, Strier has trained over forty Brazilian students, with the ultimate goal of helping them to develop self-sustaining ecology and conservation programs.[23]

COLONIAL ANTHROPOLOGY

Some areas make the transition away from the old colonial science more readily than others. Indeed, the most interesting link among the most infamous examples of colonial anthropology is that they do not end in the past but continue into the present. That is what precludes any possibility of arguing that "that was then, this is now." Rather, that was then, and it is still.

The Hottentot Venus

Sarah Baartman, also known by the Dutch diminutive "Saartjie" (pronounced "Sarky"), was born around 1790 in southern Africa to a group known as Hottentots or, now, Khoisan. After her family was killed in an ambush by Europeans, she found domestic work in Cape Town, and at the urging of Hendrik Cesars, brother of the head of household and an English doctor for whom he worked, she sailed with them to England in the expectation of wealth and love. The expectation of wealth was based on the public display of her body, especially her steatopygia (enlarged buttocks) and tablier (elongated labia), characteristic of some of the indigenous women of southern Africa. Linnaeus, who had somewhat controversially classified plants according to their sexual parts, had made a point of calling prurient attention to the pendulous breasts and tablier of "*Homo sapiens africanus niger*" in the mid-eighteenth century (he mentioned the sexual anatomy of none of the other geographic subspecies).

Arriving in 1810, Sarah caused a sensation for her appearance—she came to be known sarcastically as the "Hottentot Venus"—but as well for the question of her freedom. Slavery had recently been abolished in England, and a formal inquiry concluded that she was there of her own free will and not as a slave—in spite of the fact that the men in her life were repeatedly offering to sell her. A contemporary French cartoon ridiculed not Sarah herself but the British fascination with her—depicting a Scotsman transfixed by her genitalia as behind him a dog sniffs under his kilt, likewise transfixed.

Thinking that things might go more easily for them in Paris, Sarah's handlers brought her there, and she was received in the salons of the day before succumbing shortly to tuberculosis (a recurrent theme in these stories of bringing non-Europeans to the big city). Upon her death in 1815, she was dissected by the leading anatomists of the age, and her skeleton and genitalia were preserved at the Musée de l'Homme in Paris. She was taken off display in the 1970s, but a flurry of scholarly literature about her remains in the mid-1980s sparked a postcolonial interest in her, particularly after the fall of the apartheid South African government in 1994.

To the new South Africa, Sarah Baartman was an icon of colonialism, diaspora, and the dehumanization of African peoples as well as a symbolic ancestor. President Nelson Mandela asked his French counterpart, François Mitterrand, for assistance in having her remains repatriated to her native land, to no avail. Museums, after all, exist to collect things, not to give things away. The request was seen as a dangerous precedent, and certainly the museums of Paris had no more interest in, say, giving the Italians an excuse for demanding the *Mona Lisa* "back" than the British Museum had in honoring the very real requests of the Greek government to return the marble friezes removed from the Parthenon by Lord Elgin during Sarah Baartman's own lifetime.

South Africa had three things going in its favor, however: Baartman's remains had no scientific value, had no aesthetic value, and were just taking up space, not actually being seen or studied by anyone. After several years of negotiation between the two governments and between their two leading anthropologists (Phillip Tobias of South Africa and

Henri de Lumley of France), Sarah Baartman was finally sent home in May 2002.[24] Indeed, other requests for international repatriation are being made and acknowledged, one of the most notable being Yale's tentative agreement in 2007 to send back to Peru much of the Inca material collected by the archaeologist Hiram Bingham in the 1920s.

The New York Eskimo

Lieutenant Robert E. Peary, Arctic explorer, returned from a voyage to Greenland in September 1897 with a cargo of great scientific value, including a meteorite and six Eskimos. The Eskimos came to help organize the ethnological collection—that is to say, their stuff—for the American Museum of Natural History and would be returned to their homes a year later, where they would also, it was hoped, aid in the development of an outpost from which to launch an expedition to the North Pole. Among the six were a thirty-five-year-old man (Ke-suh, later Ke-Shu, Kushu, Kushan, Kishu, and finally Qisuk) and his nine-year-old-son Mee-ni (later, Menny, Minny, Merrie, Menee, Mene, and finally Minik). They quickly took ill. By February 1898, Qisuk was dead of tuberculosis, and four of the others soon succumbed as well. The last living adult in the group went back to Greenland, and the orphan Minik remained in New York in the care of William Wallace, a museum official.[25]

The suggestion to bring an Eskimo to New York had been made to Peary in 1895 by Franz Boas, who was at that time highly respected as an expert on Eskimos but lacked an academic post. By 1897, Boas was working at the American Museum of Natural History to organize a major expedition to the Northwest coast, and he turned the study of the Eskimo visitors over to his student Alfred Kroeber, who produced the first three publications of an illustrious career.[26]

After Qisuk's death, his body was autopsied and his brain removed and examined by the physical anthropologist Aleš Hrdlička. The corpse was then claimed for the museum by William Wallace, who prepared the skeleton for conservation, hoping to mount and display it. The *New York Times* reported, "There will be no mourning among his kinsmen, as forgetting is the Eskimo mode of dealing with the departed."[27] There were, nevertheless, funerary rituals, and it was decided to perform them for the benefit of Minik, who was still expected to return to his people and would know that the appropriate ceremonies had taken place—even if he didn't know that the deceased was not actually present.

Minik, however, chose not return to his people on Peary's next voyage but remained in New York, in the custody of Wallace. In 1901, unfortunately, Wallace was fired for financial misdeeds. In 1904 his wife died and he remarried, but his resources dwindled and he became increasingly unable to take proper care of the boy. Now calling himself Mene Peary Wallace, the New York Eskimo boy attended school in Connecticut and Public School 34 in the Bronx and excelled at baseball and ice skating. He worked briefly as "the only Eskimo theatrical usher in the world" in 1907.[28] The following year, Peary refused him passage home, for Minik apparently desired only a holiday in the Arctic and had ambitions to remain in New York as a chauffeur. Minik came down with pneumonia shortly thereafter[29] but recovered and entered Manhattan College in February 1909 to study civil engineering.

Neither (by now, Commodore) Peary nor the museum had much interest in maintaining an association with Wallace. The adopted father and Eskimo son became increasingly embittered toward both and set out to discredit them. Minik had begun to ask the museum to hand over his father's bones for a Christian burial, but their reluctance to give the bones to a teenager, for a disposition that Qisuk himself would surely not have wanted, is understandable. Minik left college after a few weeks and took off for Canada with five dollars in his pocket, mailing suicide notes along the way.

In July 1909 he was granted passage back to the Arctic by the very people he was vilifying—on the promise that he behave himself on the way and not return.[30] Minik, however, did return a few years later, offering to sell a story to the press about what the Eskimos thought about the controversy between Peary and Frederick Cook for credit in having reached the North Pole first. Peary was now a rear admiral and a national hero, and Minik encountered little interest in his story at the beginning of World War I. He worked briefly as a logger in New England and died of influenza in 1918.

The sad story of Minik is difficult to analyze because there are so many crosscutting tragedies. He was orphaned in a distant land and caught between two identities—racially othered and culturally American. But from the standpoint of colonial science the question is, Who owed what to whom?

The institution that had brought him to New York arranged for him to be adopted; sadly, his adopted father turned out to be embezzling from that institution, and it severed its connection with him. Minik's bitterness toward the museum is at least as readily explained from that indirect relationship as from its direct refusal to accede to his demand: "Give me my father's body."

Most repetitions of Minik's life follow Kenn Harper's comprehensive biography,[31] with that poignant title, and express outrage at the fake funeral for Qisuk. One could, however, alternatively see it as an act of considerable sensitivity—performing the rituals over his body that the deceased would have wanted. A Christian burial would have been inappropriate, and a long journey back to the Arctic would have been logistically impossible. And what would a teenager of very modest means possibly do with the disarticulated skeleton of an adult Eskimo?

Minik had been told by his adoptive father (who had macerated and cleaned it himself) that Qisuk's skeleton had been mounted and displayed, but that seems unlikely since the procedure would have involved drilling holes in the bones, and that was not done. Minik had opportunities to return to Greenland but declined them (being an orphan and a child, that seems understandable enough). The folks who had brought him over had set him up with a family; when that turned sour, he became another New York kid with a tough life—depressed, unstable, and angry. He freely articulated that bitterness to the New York tabloids, which went along because he had an interesting angle, but that certainly did not ingratiate him to either Peary or the museum, who were essentially his only tickets home. Even so, Minik managed to find friends of means who cared for him, and ultimately he did go home—and yet returned to New York anyway.

Normalcy as either an Inuit or an American was impossible for him to achieve. But there is an admirable end to the story. With the emerging interest in repatriation and indigenous rights in the early 1990s, the American Museum of Natural History undertook to find someone in Greenland to request the bones of the New York Eskimos, so that they could be returned. This was not mandated by NAGPRA, since it did not involve Native Americans and even predated the drawn-out negotiations between South Africa and France over Sarah Baartman's remains; it was a gesture of goodwill. Eventually the museum was successful, and the bones of the New York Eskimos were given back to their culturally affiliated relatives.[32]

All, that is, but Minik, who is buried in New Hampshire.

The Last of His Tribe

One of the participants in the funeral ceremony for Qisuk in 1898 was a student who had been collecting ethnographic information from the New York Eskimos, Alfred Kroeber. Only a few years later, Kroeber was tapped to found the anthropology program at the University of California at Berkeley and was rapidly becoming the expert on the Indians of California. In a strange turn of events, he would get to relive some of the New York tragedy in California.

Civilizing California involved the large-scale murder of the people who originally lived there. In particular, in the mid-1800s a protracted series of massacres had caused the number of Yana speakers in the Sacramento Valley to crash from around fifteen hundred to about seventy. A small band of those people, known as Yahi, retreated into hiding for the next few decades and were rarely seen, subsisting by hunting, gathering, and pilfering. By the early 1900s there were few of them left, and they were known mostly from abandoned camp sites.

On August 29, 1911, the last of them gave up. Middle-aged and nearly starving, he was backed into a corner by guard dogs as he tried to get something to eat from a meat locker in Oroville, California. He fully expected to be killed, as he had seen happen to his family and friends over the preceding decades.

He didn't know it, but an odd thing had happened to the white people in the intervening years. As they had successfully "pacified" the Indians, they no longer perceived them as threatening and had begun to romanticize them. Rather than kill the Indian, then, the local sheriff tried to talk to him and, failing that, put him in jail for his own safety and gave him food and clothing. He didn't know what else to do with the man. The local newspapers picked up the story of a "wild Indian" apprehended in Oroville and, on August 31, Kroeber telegraphed the sheriff and dispatched his protégé, Thomas T. Waterman, to take custody of the Indian. Waterman brought him back to San Francisco on September 4, and with Sam Batwi, a speaker of a related dialect, they managed to communicate with the Indian.

He came to be known as Ishi—which meant "man," since one's actual given name was highly taboo among his people, and he never actually shared it with the scientists. Kroeber, as director of the Museum of Anthropology, arranged for Ishi to have a place to live (the basement) and a job (janitor). In the age before social welfare programs, having a job and a home counted for a lot. To Kroeber and Waterman, he told his myths and lifeways; to the linguistic anthropologist Edward Sapir he just talked. To his personal physician, Saxton Pope, he taught archery, and on weekends he would give arts and crafts demonstrations to visitors at the museum. Ishi was a local celebrity, and even if the scientists wrote about him in "objective" terms they considered him their friend.

But by early 1915 he had developed the first signs of tuberculosis. Kroeber was determined that Ishi would himself dictate the terms of his departure from the mortal sphere, most especially those related to his belief in the body's physical integrity as it passes to death. He knew, however, that he would have to battle against interests that saw Ishi as a scientific specimen. In New York, as he learned of Ishi's imminent passing, Kroeber wrote to his graduate student E. W. Gifford on March 24, 1916:

> Please stand by our contingently made outline of action, and insist on it as my personal wish. There is no objection to a cast. I do not however see that an autopsy would lead to anything of consequence. . . . Please shut down on it. As to disposal of the body, I must ask you as my personal representative on the spot in this matter, to yield nothing at

> all under any circumstances. If there is any talk about the interests of science, say for me that science can go to hell. We propose to stand by our friends.

Ishi died the next day, before Kroeber's letter arrived. It probably wouldn't have mattered anyway, for then, as now, nobody listened to graduate students. He died in the hospital of the University of California of San Francisco, under the care of doctors and friends, a luxury most other victims of "consumption" did not have.

On March 30, Gifford wrote back to Kroeber:

> The only possible departures from your request lie in the fact that an autopsy was performed and that the brain was preserved. However, the matter, as you well know, was not entirely in my hands, as I am not the acting head of the department. In short, what happened amounts to a compromise between science and sentiment, with myself on the side of sentiment. Everything else was carried out as you would have done it yourself, I firmly believe. The Indian told [Saxton] Pope some time ago that the way to dispose of the dead was to burn them, so we undoubtedly followed his wishes in that matter. In the coffin were placed one of his bows, five pieces of dentalium, a box full of shell bead money which he had saved, a purse full of tobacco, three rings, and some obsidian flakes, all of which we felt sure would be in accord with Ishi's wishes.

The autopsy, while against the wishes of Ishi and Kroeber, was nevertheless standard procedure for deaths at the hospital. The removal of the brain was not.

Kroeber's second wife, Theodora ("Krakie") Kroeber (his first had died of tuberculosis a few years before Ishi), published *Ishi in Two Worlds* in 1961, and it became an anthropological classic, bringing Ishi's story to generations of readers, including many high school students in California. She included the bit from Gifford's letter of March 30, 1916, about the brain being preserved but said nothing else about it. A few decades later, after the passage of NAGPRA, a group of four Native American tribes from northern California, knowing that Ishi's brain had been "preserved," set out to reclaim it as a symbol and as a relative's body part. The Butte County Native American Cultural Committee, led by Art Angle, made inquiries and followed up rumors about where the brain was, but they got nowhere. So they went to the *Los Angeles Times*, which wrote up the story of Ishi's missing brain on June 8, 1997.[33]

Now the UCSF Medical School initiated its own investigation, feeling a degree of responsibility because the procedures had been done under its auspices all those years ago. The vice-chancellor for academic affairs assigned the medical historian Nancy Rockafellar the job of investigating what actually became of Ishi's brain. But Rockafellar quickly ran into a brick wall as well. Told that the brain had been sent to the Smithsonian Institution in Washington, D.C., she contacted the Smithsonian and was told that they didn't have it and that it didn't exist.

A Duke University anthropologist named Orin Starn, however, was going through Kroeber's extensive correspondence and was able to piece together the missing parts of the story. Kroeber had returned from New York to find that his friend had been cremated and interred, but his brain had been omitted and Kroeber was now stuck with it. What to do with your Indian friend's pickled brain? Kroeber wrote on October 27, 1916, to the Smithsonian's physical anthropologist, Aleš Hrdlička, who had studied Qisuk's brain, and offered the new one for his collection. Hrdlička gladly accepted. Kroeber shipped it off on January 15, 1917, and it was given accession number 60884, museum number 298736. It was moved in 1981 and again in 1994. But it was pretty clearly last seen in the possession of the Smithsonian.

Armed with the relevant correspondence, Starn drove to the Smithsonian and met with officials on January 27, 1999, who now acknowledged having the brain (Oh, *that* brain!) and began to consider the request for repatriation. But they moved a bit too slowly for the state legislature of California, which held a hearing on April 5, 1999. The anthropologist Nancy Scheper-Hughes read into the record an apologetic statement from Berkeley's anthropology department that encouraged the rapid transfer of the contested cranial organ. The Smithsonian adopted the position that had been working for many museums: how do we know who the rightful owners of the brain are? A reasonable question, to be sure, but also a useful stratagem for maintaining control of the materials more or less indefinitely. In this case, however,

California produced representatives from Native American tribes all over northern California to say that they supported the Butte County Coalition's repatriation request—and gave the Smithsonian a month to do something about it.

There was one final opportunity to perform a positive and healing act. On May 7, the Smithsonian announced that it was returning Ishi's brain, but . . . not to the troublemakers of the Butte County Coalition; it was instead giving it to two other tribes that spoke languages related to Ishi's Yana. In other words, the museum would maintain its control, up to the bitter end, over the brain that it had denied having in the first place, and it would make sure that the folks who had gotten it into this mess didn't come away with what they wanted. But they did, for it was not so much an issue of material possession to the Indians but of symbolic unification.

More than a year later, in August 2000, Ishi was reunited with his brain and reburied—by those who got him, together with those who had asked for him—in a secret place in northern California. Shortly thereafter, the Pit River tribe and the Redding Rancheria—now the lawful custodians of the reburied brain—hosted a remembrance ceremony for Ishi and invited representatives from Berkeley and the Smithsonian and those who had helped bring the matter to a fairly happy conclusion.[34]

SCIENCE AND PEOPLE

What the previous stories share is a beginning in premodern scientific sensibilities and an ending—often a tortuous one—in a very different set of ideas about the relationships between science and its human subjects and about the responsibilities of the former toward the latter.

The central question of colonial science was, What can you get away with on other people that you can't get away with on your own? The other people were usually foreign or different and always socially inferior, or subaltern, and this is clearly the kind of thinking that went into the (now) notorious Tuskegee experiments, on American soil. In

Alabama in the 1930s, doctors thought it would be interesting to study the physiological effects of advanced syphilis. But they rarely got to see such a stage, and when they did they felt an obligation to treat it, although the treatment was difficult and dangerous. So they resolved to identify a study population—poor black men—who would come to them infected with the disease and receive medical attention, but not treatment, for it. The U.S. Public Health Service workers rationalized it with the knowledge that, after all, they didn't actually give these guys the disease—they just didn't try to cure it. In that sense they could feel as though they were following the letter of the Hippocratic oath, which begins, "First, do no harm"—if not its spirit.

The study stretched for decades, with the involvement of black health care workers to recruit subjects for the study, through the Nuremberg trials of Nazi scientists, through the development of penicillin as an effective cure that was denied the Tuskegee participants, through the unease and ignored reports of potential whistle-blowers, and into the 1970s. The story finally broke in July 1972, and the study was ended—but not until after men had died as a result of not being treated, spouses and others had been infected, children had been born with congenital syphilis, and obviously many people had suffered needlessly as a result of the callousness of the science. President Bill Clinton offered a public apology to the survivors in 1997.[35]

The subaltern subjects may be even closer to home, however. In 1997, I visited the South African National Museum and saw a wonderful exhibition called "MisCast" on the way the indigenous peoples had been represented, studied, and used—including, of course, Sarah Baartman (pre-repatriation). One thing that caught my interest was the old photos of "Bushmen" posing naked for scientists. The text was minimal, for the degradation of having naked photos of yourself taken for science seemed self-evident.[36]

But it wasn't quite so self-evidently degrading. In this case, scientists were doing to the Khoisan just what they were doing to their own teenage children at Yale, in what had recently come to be known as "The Great Ivy League Nude Posture Photo Scandal." In 1992, Naomi Wolf, author of the best seller *The Beauty Myth,* gave a commencement speech at Scripps College that was published shortly thereafter in the *New York Times*. In it, she took the decidedly un-macho television host Dick Cavett to task for his joke at her own graduation back in 1984, concerning some naked photos of Vassar undergraduates (and how ugly they were). Cavett responded indignantly that his comments about the photos had been evenhanded and had come directly after a reference to "Yale's silly male posture pix of yore." Moreover, added the Yale art historian George Hersey, the photos in question were taken at many U.S. institutions of higher education and were not so much pornographic as ostensibly scientific. Finally, in 1995, Ron Rosenbaum published an article in *New York Times Magazine* that explained what it was all about.[37]

In the early decades of the twentieth century, America's elite colleges were (as they still are, although with less of a monopoly) producing the country's leaders. They were not merely educational institutions but finishing schools as well: they made leaders out of teenagers. Obviously you needed good posture to lead; who would follow a slouch? Out of such considerations was born the idea of taking a standardized photo of incoming frosh to determine who might need a bit of remedial uprightening. The process began at Harvard and Yale, and other colleges soon followed suit, naturally trying to emulate Harvard and Yale. Sometime in the 1930s, however, the process evolved from merely photos of posture to bits of scientific data—that is to say, photos of the student body's bodies. And with that, the clothing somehow disappeared as well.

The transformation from helping students look more like leaders to the creation of a scientific database of the bodies of naked eighteen-year-olds was organized principally by William H. Sheldon, a psychologist-physiologist and protégé of the Harvard physical anthropologist Earnest Hooton. Indeed, Sheldon's pompous descriptive terminology (*mesomorphic* for muscular, *ectomorphic* for skinny, and *endomorphic* for fat) remains widely in use. His goal, however, was something grander: to associate body form with personality, temperament, and ability. At one level, the endeavor was trivial; you're never going to see a short, fat guy playing shortstop in the major leagues unless he goes on a crash diet and gets very light on his feet. But that was not what Sheldon and Hooton were after; they wanted to know what the innate, hard-wired components *below* the neck were, which they believed to be biologically and deterministically coincident with the innate, hard-wired features *above* the neck. Not surprisingly, they never got beyond generalizations like fat people are jolly and skinny people are conspiratorial.

Nevertheless, for over four decades, eighteen-year-olds at America's elite colleges were put through an orientation routine that went pretty much like this: on your first day, you will awake at 8 A.M., then shower and groom yourselves, have breakfast in the dining halls, pick up your class cards, return to the dining halls for lunch, go to the gym and strip naked and pose for a set of full-body pictures for us, then join some intramural clubs, speak with your faculty advisor, and return to the dining halls for dinner.

And for nearly half a century, nobody objected. Why? Well, this is the way we do things at America's elite colleges. Who are you to question it? You know who makes trouble? Communists! Anarchists! That's who! You're not one of them, are you? No, I didn't think so. Now go over there and take your clothes off, so we can have a snapshot of you for our files.

The bottom line is this: if that's the way they treated the future captains of industry, it can hardly be surprising that the Khoisan were treated similarly.

The posture photos ended in the 1960s, when many of the schools went co-ed, and students actually didn't mind being called communists or anarchists anymore. But when Ron Rosenbaum's article came out in 1995, it raised questions about responsibility. The photos were presumably anonymous, and the faces were obliterated—yet there were posture photos somewhere of many of America's current leaders. Yale had turned most of its photos over to the Smithsonian and now asked the Smithsonian to destroy them, since they seemingly constituted a set of scientific data collected under coercive conditions. There was still debate about this at the business meeting of the American Association of Physical Anthropologists. I can remember a senior scientist standing up to decry the destruction of data, to which I responded that if anyone could design a scholarly project that might use thousands of photos of naked rich white eighteen-year-olds, devoid of any context or associated information, I'd be for preserving them, but I sure couldn't think of one. (Over lunch one day, George Hersey corrected me: The photos might indeed have use to an art historian, even if not to a scientist!)[38]

The biggest irony is that of Naomi Wolf, Dick Cavett, George Hersey, and Ron Rosenbaum—the ones who had put the whole ordeal into the public eye—only Wolf herself had been spared the humiliating experience of the Naked Posture Photo, for the project had ended years before she entered the Ivy League.

GENETICS AND THE DAWN OF HEMO-TOURISM

In the early 1990s a group of population geneticists, led by Stanford's Luca Cavalli-Sforza, proposed an ambitious scheme to the scientific community and to the public. They wanted to take blood samples from the indigenous peoples of the world—tallying about eight hundred such populations—to bring back to Palo Alto, California, and be studied. Why? They wanted to know who was closely related to whom and try to work out the microevolutionary history of the human species.

An interesting question, to be sure, but not particularly resonant with the people whose blood they wanted. Why should anybody want to help geneticists delegitimize their own beliefs about who they are and where they came from? And who knows what other studies would be done with the DNA; would geneticists then use the same samples to try to prove that the indigenous people are mentally deficient, as Bruce Lahn (chapter 6) attempted to show? To be sure, collecting blood from the field was an age-old practice of biological anthropology, but always on a small scale. On a large scale, it was raising questions that had previously been avoided, and the payoffs for participants in this research did not seem particularly great. So the geneticists began groping for reasons to overcome reasonable questions about the project—such as local taboos about blood and body parts, bioethical concerns about the nature of informed consent in a cross-cultural setting, and the increasing cash value of cell lines and exotic bits of DNA to biotechnology. They began to wax prosaic about curing genetic disease and refuting racism scientifically, neither of which was realistic, when all they really needed to do was show that the social and bioethical issues mattered, and that the scientists were willing to work toward coping with them.[39]

Ultimately, in the late 1990s, the Human Genome Diversity Project was denied the federal funding it sought and quietly gave up the ghost. Its principal accomplishment was to have gotten Native American groups actively antagonistic about the field of human population genetics.

In 2005, however, the human population geneticists were back on the scene, this time as the Genographic Project, led by Spencer Wells. This project had the same goals as the original Human Genome Diversity Project, but with one major difference. It had its own funding, principally from National Geographic and IBM. "Private funding" means something critical in practice, however—namely, that the researchers no longer have to worry about the proprieties involved in acquiring DNA samples from indigenous peoples. In other words, ethics, schmethics—the world is simply obliged to take them at their word, that their intentions are noble and their deeds will be nice and benign. The Genographic Project made a public show of going through its researchers' universities' institutional review boards—but like the aliens in "To Serve Man," the classic *Twilight Zone* episode ("It's a cookbook!"), the sincerity of that act was quickly called into question.

A Genographic Project researcher received clearance from his university's board to collect some DNA samples from Native Alaskans, pending the approval of a local Alaskan review board. He didn't get the local approval but collected the samples anyway. Understandably, when word got out, the Alaskans angrily demanded their DNA back. The Genographic Project spent the next year or so negotiating to try and keep the samples they were not supposed to have collected in the first place.[40] Not a promising start.

In 2007, the project began to solicit wealthy patrons who might want to go on the "Journey of Man by Private Jet." For $50,000, you could visit exotic subaltern people from Mongolia through New Guinea and the Kalahari Desert and actually have your own DNA tested so that you could pretend to be their kin—and all in the comfort of a customized Boeing 757. If you wanted a single room, however, it would be another

$7,500, and you would still be hit up for an additional $2,000 to help support the Genographic Project, *because it is such a good cause!*[41]

This marked the advent of a new kind of colonial science: hemo-tourism. It seems as though the apprehensions of indigenous people being exploited by geneticists were well founded after all. Now, that's ironic: sometimes even laboratory rats learn faster than scientists.

NINE Racial and Gendered Science

To an outsider like me, who doesn't know him, James Watson is an enigma. In 1953, with Francis Crick, he reasoned out the structure of DNA and literally invented the field of molecular genetics. In 1962 he won the Nobel Prize for it. His 1968 memoir *The Double Helix* is a foundational work of science studies, since it was really the first "ethnography" of laboratory life. And it caused a sensation for its candor—even if it was a firsthand, self-interested account—in describing the researchers as egotistical cutthroats rather than as the humble, social altruists that scientists generally wanted to be seen as. Somewhat later, as head of the Human Genome Project, he created a large pot of money—4 percent of the project's budget—for Ethical, Legal, and Social Implications, although he also let it be known that he didn't want this part of the project to do anything and hoped it would be run by Shirley Temple.[1]

Stumping for the Human Genome Project, he told *Time* magazine in 1989, "We used to think our fate was in the stars. Now we know, in large measure, our fate is in our genes."[2] That popped some eyes open and was probably the most deconstructed scientific utterance of the decade. Did Watson really think that genetics was like astrology, only presumably more accurate? Did he really think that there is "fate," in any widely understood sense of the term? And if there is, did he really think that it has been localized to our cellular nuclei? Or was he just talking bullshit to the public in order to get the federal funding he was after?[3] Either way, it was pretty weird.

At a lecture in Berkeley in 2000, Watson told the stunned audience that there was a deterministic link between libido and skin color. Of course, he wasn't reporting on a new discovery; he was just pretty sure it was there. A former colleague's best-selling memoir gave him an unflattering epithet: "the Caligula of biology."[4]

Finally, in October 2007, Watson was on a publicity tour for a new book and gave an interview to a former student, now writing for the *Times* of London:

> [Watson] says that he is "inherently gloomy about the prospect of Africa" because "all our social policies are based on the fact that their intelligence is the same as ours—whereas all the testing says not really." . . . His hope is that everyone is equal, but he counters that "people who have to deal with black employees find this not true." . . . He writes that "there is no firm reason to anticipate that the intellectual capacities of peoples geographically separated in their evolution should prove to have evolved identically. Our wanting to reserve equal powers of reason as some universal heritage of humanity will not be enough to make it so."[5]

Right-wing Web sites posted it immediately. Watson quickly recanted the statement, but he couldn't retract it. His book tour in Great Britain was canceled, and he was suspended from the Cold Spring Harbor Laboratory he had been running since 1968 (coincidentally the very institution started up by the racist eugenicist Charles Davenport decades earlier).

And yet, there were contrary voices. Does he not have a right to his opinion? Aren't there scientific data to back up his position, even if it is politically incorrect? Ah, here we get to heart of the matter. There are data out there to support all kinds of bunk. Crop circles, cattle mutilations, hypnotic memories, and eyewitness accounts all attest to extraterrestrial visitations to earth, but no serious scientist takes them at face value. It's not just data but the ability to evaluate data critically and make sense of the whole picture that produces a scholarly scientific account. Only a complete ignoramus takes all data at face value. That is why Steven Rose, the distinguished British biologist and social activist, weighed in strongly: "If [Watson] knew the literature in the subject he would know he was out of his depth scientifically, quite apart from socially and politically."[6]

So the issue is not whether the scientific book is still open on the innate intelligence of Africans vis-à-vis Europeans, but whether the person making the claim knows what on earth he's talking about. Of course, if you're a Nobel laureate, the presumption is that you do know what you're talking about, because you are really, really smart. But sometimes being smart can entail shutting up and not exposing the scope of your ignorance, so that your prejudices are not confused with actual knowledge.

Indeed, it has already been shown that a Nobel laureate and a racist ignoramus can inhabit the same body. William Shockley won the Nobel Prize in physics in 1956 for inventing the transistor and essentially founded Silicon Valley. But in the face of the social unrest of the 1960s, he became convinced—like Madison Grant before him—that social problems were the result of the proliferation of genetically stupid people, that genetic stupidity was not evenly distributed over the world's populations, and that a program of sterilization was the solution.[7]

So what are the problems here?

First, the Nobel laureate thinks he is a trailblazer, but that is only because he is woefully unaware of the history of the subject. There isn't a scientific endeavor in which you could reasonably expect to make a contribution without a minimal familiarity with what has already been accomplished, what is already known, and what is most likely to be a false and misleading trail. Shockley himself successfully sued a newspaper that had called him a Nazi—but he won a judgment of only one dollar.[8]

Second, we know a little about the genetics of intelligence, don't we? We know about the cultural limitations of intelligence tests; we know about the effects of social rank on performance; we know about the equilibrating effects of income, family structure, parental education, civic expenditure on public education, health, nutrition, motivation, self-image, and general expectations—and of the consequent necessity to control carefully for those variables. We know that when those variables are not controlled there is a fifteen-point average difference between U.S. blacks and whites, and when those variables are controlled there isn't. We also know that when these variables are poorly or incompletely controlled, as in the 1994 best seller *The Bell Curve*, there is a small difference—and that it is far more likely to be the result of the real statistical treatment than of imaginary hereditary defects.[9]

And any global traveler knows that poorly schooled Africans can nevertheless commonly speak five languages fluently, and that poorly schooled street vendors can cheat you in four interconvertible currencies. That sounds pretty intelligent, but it is not incorporated into a linear test score.

Third, we know how poorly patterns of genetic diversity map onto the categories of people being compared. We know that people are genetically different, that some people can do things better than other people, and that groups of people have shown prominence in different areas. The first two, however, incorporate the genetic differences of people *within* a group, while the third is about *group-level* differences. We also know that the causes of diversity within a group cannot be easily extrapolated to the causes of diversity between groups.

Finally, we know that pathology does not explain normalcy; the leading genetic causes of mental retardation (e.g., fragile-X and Down syndrome) have nothing whatsoever to do with the vast numbers of people who come out disappointed with their scores on pencil-and-paper tests.[10]

Why, then, is a Nobel laureate molecular geneticist out there making racist claims, which he may or may not believe, to the press? Is he a racist or an ignoramus?

My opinion—and as I said, I've never met him—is that he is neither, but that he is even more loathsome, especially for someone in his position—that of an authority and a scientist in the public eye. And I think the clue lies in his 1989 comment that "our fate lies in our genes."

That clue isn't part of a standard science lesson, but it is probably more important: when someone very smart says something apparently very stupid, there is generally a reason for it. We all feel the tug between doing what's right and doing what's expedient; between saying things that the listeners want to hear and what they need to hear; between maintaining a sense of dignity and self-respect at some personal cost and being for sale to the highest bidder. Some people set the bar far to one side; others set it far to the other side. At one end you have Ayn Rand's smug architect Howard Roark, who believes he's a genius and consequently never listens to anybody else, for that would compromise his integrity. At the other end you have the whore or junkie who will do absolutely anything to anyone for five bucks. I think most of us are somewhere in the middle, weighing the means against ends, the costs against the benefits, what stands to be gained against what we can live with in order to achieve it. What we differ in is where we put the bar.

In this framework, here is how I think of Watson: I think he's the junkie.

In 1989, Watson was trying to drum up support for a federal investment of $3 billion in a scientific program to sequence the human genome. This came just at the time when popular interest was turning away from the multi-billion-dollar investment in the Superconducting Super Collider—a project that would employ engineers and physicists until the Second Coming and tell us more than we could possibly imagine about subatomic particles like the Higgs boson.

But the reasonable question asked by the American public was, Why should we give a fuck about the Higgs boson?

Physicists found that question surprisingly difficult to answer. *They* cared about the Higgs boson, so we should too. It might tell us about the structure of matter, after all. It might tell us about the origin of the universe. Um, maybe we could use that knowledge to build a time machine. What are you, anyway, anti-science?

No, I just think there might be better ways to spend a few billion dollars. Developing ways to break our dependence on fossil fuels, feeding the poor, raising literacy rates, fixing roads and bridges, improving public schools, that sort of thing.

It was into this public discourse that the Human Genome Project was introduced. What it needed to show, above all, was that sequencing the human genome was a good way to spend several billion dollars. In particular, it needed to convince the public that this was the most important scientific information you could ever want. It was science that was vital. That's why the popular science at the time became flooded with some of the most purple genetic-determinist prose ever devised: the code of life, the book of man, the human blueprint.

And that is what Watson was saying when he told *Time* magazine, "Our fate is in our genes"—yes, this is the greatest financial investment you will ever make. It is the most important scientific project ever conceived. It will be the best $3 billion you ever spent. Just give me the money and stand back. And the proof of the pudding, so to speak, is that by 1993 the Superconducting Super Collider was dead but the Human Genome Project was alive.

The lesson learned, however, was this: say anything at all to the public to convince them to keep funding levels high for molecular genetics. If racial differences in intelligence genes come to mind, however vulgar it may sound, and however anti-scientific it may really be, it can and should be put out there. You don't necessarily have to believe it (although that helps). The most important thing is just to keep the spigot turned on; and you say or do anything to get that cash. That's why I don't think Watson's comment was either self-consciously racist or completely ignorant. I don't even think it was necessarily about race at all—I think it was instrumental, and about business, and it followed a strategy that had been successful for the field for a couple of decades.

THE TWO GENETIC LAWS

Geneticists like to teach about two laws—Mendel's first and second—in spite of the fact that (1) Mendel, who peaked in 1866, never formulated them; (2) nobody even thought about separating them into two laws until 1916; and (3) they strictly apply only to humans in rare and aberrant cases, and then for bizarre things like earwax texture, earlobe shape, the ability to roll your tongue, and your blood group. Consequently, their strict application to humans is really rather limited, unless you happen to think that Neandertals died out on account of their earlobes. Even then, the ability to roll your tongue isn't really a Mendelian trait (this idea was retracted by the geneticist Alfred P. Sturtevant shortly after he suggested it, but it had such great heuristic value that it continues to be taught), and the ABO blood group has three major alleles, two of which are codominant, which is a situation far more complicated than Mendel even dreamed.

Here are two laws about human genetics that are actually more generally applicable than Mendel's laws. First, *99 percent of everything a geneticist says publicly is code for "give me more money."* Once you internalize that law, the literature on human genetics becomes much more understandable. Second, *geneticists always talk about the future as they reach for your wallet.*

Let us take them in turn. The First Law explains James Watson's apparent racism, or at least his convergence with racist ideologues, who quickly brandished his words as if they were engraved on flaming crosses. It explains why the gene mapping literature is so replete with "genes for things" that have been mapped and unmapped with appalling regularity—the mapping being big news and the retraction or failure to confirm being quietly buried. It explains why early Americans first crossed the Bering Strait thirteen thousand years ago and fifty thousand years ago, in one wave, two waves, three waves, and in a continuous stream. It also explains the nearly ubiquitous appearance of sentences like these in the relevant literature (with my responses to them thrown in for good measure):

- These results are preliminary. (Then you shouldn't have published them, should you?)
- Further studies are needed. (Sure, as long as you're not the one who does them.)
- Future research may confirm these results. (But probably won't.)
- Our calculations show that this species diverged from the others longer ago, or more recently, than anyone else believes. (Then go home and do some better calculations.)
- These results on sexuality in hermaphroditic worms may shed some light on human sexuality. (*Man, you are one sick puppy.*)[11]

The Second Law is a bit slimier, but a logical consequence of the first. If geneticists make extravagant claims in order to keep public interest aroused and funding intact, then at some point they have to deliver. This is not to say that no good has come of genetics—certainly genetic screening, lactose-free milk, and mice that glow in the dark are all salutary consequences of genetic research. But they are the products of more or less normal science and came about independently of the Big Science of the Human Genome Project. What hath the Human Genome Project actually wrought?

The anthropologist Mike Fortun explores the relationship between genetics and futurology. Geneticists are always talking about what they're going to do, but they never actually seem to get around to doing it.[12] I was at a conference not too long ago at which that question was addressed by a molecular geneticist, with utter condescension. He had asked the audience whether anyone disagreed with the statement that the $3 billion spent to produce the human genome sequence "has been good value for money"—and was astonished when someone (well, me) raised a hand. After all, I was old enough to remember the promise on which the expenditure was principally based—that we were going to cure genetic disease with the human genome sequence. But, as of today, no genetic diseases have been cured—since there is no gene therapy, nor is there likely to be in the foreseeable future. It's killed a few people but hasn't really cured anyone. So how on earth could any sensible person consider that good value for money?

The geneticist responded that someday soon we may well be able to cure genetic disease. I said that the original question was posed in the present perfect but that he was answering it in a different tense. The question he posed wasn't whether it might ever turn out to be good value for money but whether the $3 billion *has been* good value for money. Someone else piped up that we've learned that there are only twenty-five thousand genes. I responded that I could have told him that for only $500 million.

The geneticist got the last word in: Come back in ten years, and you'll see how it's been good value for money.

And, by the way, Give me more money.

THE GENE AS CULTURAL ICON

The real legacy of the Human Genome Project has been not so much biological or medical as cultural. That cultural legacy even has a name: *genohype*—the exaggerated claims made by geneticists about what they have done and what they are going to do in order to keep the cash flowing.[13] To the extent that the public gets its knowledge of science from the experts, or filtered through journalists who themselves get it from the experts, it is no surprise when people actually start believing it. It means the genohype has done its job.

A eugenicist at the University of Chicago in 1929, sounding almost like the Human Genome Project several decades later, told the press that a careful selection of mates—that is to say, eugenics—could wipe out cancer in two years. Just give us the money, obey us ruthlessly, and stand back—and cancer will be gone. Oh sure, if anyone in your family has ever had cancer, we won't let you get married and have a family—but that's the science you're paying for. And if cancer isn't gone in a couple of years, it'll be your own fault.[14]

And there's also the little problem about running other people's lives. Scratch a geneticist and find a bit of a fascist, perhaps? Watson again: "People say it would be terrible if we made all girls pretty. I think it would be great." Watson was presumably thinking hardest about the word *pretty*, but others focused on the word *made*. We would, after all, be forced to abide by his taste in beauty, not to mention his decisions about marriage, breeding, and medical intervention. That is what it would take for a geneticist to make "all girls pretty."

But, you protest, someone would have to have a monstrously large ego even to entertain such thoughts about running other people's lives and being the final arbiter of all aesthetic judgments. And then you learn that the first diploid human genome produced, which belonged to a real live person (as opposed to the haploid hodgepodge announced with such fanfare in 2000), was not Nelson Mandela's, or Katrina vanden Heuvel's, or Muhammad Ali's, or Stephen Sondheim's, or David Beckham's, or even Victoria Beckham's. It was . . . James Watson's. The second real genome belonged to his rival, the genomics entrepreneur

Craig Venter.[15] The lesson: don't underestimate the monstrosity of the egotism involved.

Another eugenic enthusiast waxed eloquently: "Unless a man understands heredity he can not possibly understand human life. He can not understand one of the largest forces—probably the largest—that has made him the kind of mortal being that he is."[16] Sounds a bit like James Watson too, but once again, written four years before Watson was even born, by a science evangelist named A. E. Wiggam.

What was it all about? Getting the public interested in genetics by overselling its importance, its influence, its meaning, and its role. Then, on the side, there's a bit of plausible deniability: "Oh, he's just a *popularizer*—no real scientist takes him seriously." But in fact they *did* take him seriously, by allowing him to carry on, because what he was saying *was good for business.* Wiggam was also on the American Eugenics Society's advisory board, alongside all of those credentialed geneticists.

The eugenics movement crashed with the stock market and then the Nazis. The Human Genome Project, however, half a century later, rediscovered the formula for its transitory success: keep telling people how important genetics is, keep the drums beating, the news rolling in, the masses in the dark, and the critics demonized. Genetics is science, and thus to be against what a geneticist says is to be against science. That argument may be wrong and stupid, but it has a lot of rhetorical force.

So the legacy of the Human Genome Project, twenty years after its inception, has been that more and more people think that genetics can answer all the questions they can frame about human life, nature, and destiny. The right-wing bloggers love it and can readily dismiss their opponents not only as "politically correct" commies but as anti-science to boot.

And yet there is much more to the picture than simply pro-science and anti-science, pro-genetics and anti-genetics, pro-racism and anti-racism. Somewhere in there, a critical look at just what the science is, and where it ends, has to be undertaken. As the sociologist Dorothy Nelkin and the historian Susan Lindee explained, the gene has become not just an eponymous unit of heredity but a fetish object—a cultural icon.

Genes in fact mean different things to different groups of scientists who work on them. To one, a gene is a unit of information—an instruction; to another, it is a unit of transmission—a packet; to another, it is a unit of structure—a brick; and to yet another, it is a unit of replication—a copy. Sometimes a gene is taken to be a predictor of a physical attribute, and sometimes it is a cryptic part of a developmental pathway.[17]

So what do we understand when we hear about a "gene" for sickle cell anemia or novelty seeking, or a "genetic basis" for differences in height or intelligence, or a "genetic predisposition" for cancer or polygamy, or that bipedality and rape are "in our genes"? The answer is that we hear many things, and somehow geneticists are only rarely very keen to set the record straight about them. That task, then, falls to people who have to criticize the geneticists for their opacity in relating their work meaningfully and realistically to human experience.[18]

RACE, ETHNICITY, AND ANCESTRY

One of the greatest misunderstandings of contemporary science is the idea that geneticists have anything reasonably authoritative to say about race. Now, we all know what race is, don't we? It is a fairly large and fairly natural division of the human species. By "fairly large" we mean that there are rather few of these divisions, and by "fairly natural" we mean that they are generally homogeneous and distinct from one another, so that most people can be unambiguously assigned to one race or another.

These criteria have been applied in three different ways over the past century, however. In the early 1900s, your race was something that was diagnosable, for it resided within you. The scourge of race scientists was someone pretending to partake of a race that was not really their own—that is to say, "passing." In this view, then, a race was an abstraction; as an abstraction, it could be assigned to groups of diverse sorts and identified within their members: Blacks, Jews, Gypsies, Nordics, Celts, Poles.

A Russian hematologist actually developed a test in the 1920s that could distinguish the blood of Russians from Jews and of Poles from

Latvians, by adding a few simple chemicals, shaking, and observing what color it turned. The test also worked to distinguish the blood of men from women, male plants from female plants (in spite of the difficulty in obtaining blood from them), and homosexuals from heterosexuals. Of course, the test was rubbish, but to this day we don't know exactly what was wrong with it.[19] Suffice it to say that, when you know what answers you want, it isn't that difficult to find them.

From the mid-1930s onward, however, a crucial revision was occurring to the race concept. Instead of being something within you, that is to say, something fundamentally metaphysical, a race instead became something you were in. In other words, it became a geographically localized group of people of which you were a member—a population. This made it more amenable to genetic analysis, but the genetic analysis quickly produced a quandary: Each population could be genetically distinguished, but that didn't tell you how to aggregate them into races; the decision about whether Europeans came in one flavor (vanilla), three flavors (Nordic, Alpine, and Mediterranean), or ten flavors was essentially arbitrary.[20] If you wanted to contrast, say, Europeans against Africans, you were *assuming* that the categories were real and natural, not discovering it.

Consequently, genetics was no better at discerning natural groups of people than classical eyeballing was. The geneticist William C. Boyd attempted to do this in *Science* in 1963 and identified thirteen races: Basques, Lapps, Northwest Europeans, Eastern/Central Europeans, Mediterraneans, Africans, Asians, Indo-Dravidians, American Indians, Indonesians, Melanesians, Polynesians, and Australians.[21] The cultural ideology implicit in identifying one kind of African but five kinds of Europeans did not seem to register with him.

Another change to the concept of race, however, was already under way. It was increasingly becoming clarified that the human species was not divisible so much into biological units as into biocultural units. Our species was constituted by "a widespread network of more-or-less interrelated, ecologically adapted and functional entities," in the words of the Oxford physical anthropologist Joseph Weiner. Moreover, when it came to higher-order clusters, these were arbitrary and ephemeral.[22]

The judicious use of genetics showed that most of the detectable variation was found within any group (polymorphism)—that is to say, to a first approximation, there are all kinds of people everywhere. It also showed that the primary determinant of genetic difference between two groups of people was simply geography—how far apart they lived. But, once again, that didn't tell you whether there were three, twelve, or sixty kinds of people.

Moreover, it was already clear as well that many of the features that differentiate human groups from one another—their patterns of speech, behaviors, modes of dress, body language, life expectations, food preferences, even characteristic smells—were not really consequences of their gene pools but of something else that needed to be separated analytically from their "natural" distinctions. These came to be grouped as *ethnicity*.

Thus, by the 1980s a consensus had been reached that races were not accessible to the geneticist because they were not genetically real entities. They were very real, of course, as lived experiences in particular social and political circumstances. But from the standpoint of biology, races were arbitrary aggregates of bioculturally constructed local populations. And the biological distinctions between populations were very easy to overestimate.

So what is race? Race is an active negotiation between *difference* (which may be biological and objectively measured) and *otherness* (which is subjective and arbitrary). Race involves culturally deciding that some differences are not important and submerging them and deciding that others are important and exaggerating them.

Further, to the extent that we can analytically separate and compartmentalize these, most differences between groups of people are cultural—that is to say, are not genetically based. What isn't cultural is for the most part polymorphic—that is to say, present in most or all populations in differing proportions. What isn't cultural or polymorphic is mostly clinal—that is to say, differing across human populations gradually as geographic gradients. And what's left, that isn't cultural, polymorphic, or clinal, is principally local. That is how humans vary.[23]

But genohype gave race one last gasp. Surely we can study the differences among West Africans, East Asians, and northern Europeans? (Forensic experts had been doing that for years.) Aren't those racial differences?

Well, no. Race is a theory of classification, and from its initial use in the eighteenth century up to its scientific death throes in the 1960s it has been considered a taxonomic category equivalent to the subspecies.[24] We already know that geography is the principal determinant of difference. The fact that you can contrast the most extreme peoples against one another says nothing at all about the existence of categories or the overall structure of human variation. All Africans aren't West Africans: the continent encompasses the tallest and shortest peoples in the world and peoples of diverse facial form and pigmentation. The earliest anthropological fieldwork had shown that.[25]

Knowing about the gene pool of West Africans may mislead you about southern Africans (who have much less risk of sickle cell anemia), and knowing about northern Europeans may mislead you about southern Europeans (who have lower risk of cystic fibrosis and higher risk of thalassemia and lactose intolerance). This contrast, then, is not about race, as the word is generally understood, but about the differences among the most diverse peoples, who are not representatives but extremes.

But the conjunction of genohype and venture capital has produced a new kind of service. Contrasting a few dozen cell lines derived from northern Europeans, East Asians, and West Africans, you can identify the genetic regions where they differ the most from one another. Then you can take an unknown sample (say, a client's) and observe its pattern of resemblance to the differences you have established among the three groups of cell lines. From there, you judge that your client's sample resembles the East Asian panel in some ways, the northern European in some ways, and the West African in some ways. All you need is a statistical algorithm to condense that information into a simple, presentable framework, and then you can tell people that they are—say, 58 percent African, 11 percent European, and 31 percent Asian.[26]

And you know what? They'll buy it.

Taking such calculations at face value means imagining a historical model of archaic pure races who lived at the margins of the Old World and subsequently commingled to produce modern peoples. That was actually a nineteenth-century gloss on the fates of the biblical sons of Noah, who spread out to the corners of the earth and then became fruitful and multiplied—but that model intersects with reality at no point. The oldest human populations we know of are the ones right there in the middle—East Africa and Southwest Asia.

The other contribution of genohype to ancestry is the business that involves identifying your mitochondrial DNA lineage and matching it to your long-lost ancestors and relatives. This is being aggressively marketed in the United States at African Americans, who are often curious about their erased African roots. It is a potentially very lucrative market, and it certainly looks like science. Well, it *is* science. It's done by scientists, and it's done on DNA samples. And it produces real data.

There are only two problems with it. One is that the match is nonspecific. They tell you that you match a member of tribe A and invite you to think that you therefore don't match tribe B next door, when in fact you simply match a *sample* from tribe A and would also match a sample from tribe B if they had any samples from tribe B, which they don't. Human genetic variation is largely polymorphic and clinal, after all.

Second, they invite you to imagine your mitochondrial ancestor as your unique ancestor. But the fact is that, as recently as three hundred years ago, your direct ancestors numbered well into the thousands, yet your mitochondrial DNA identifies only one. That is to say, far fewer than one-tenth of 1 percent of your ancestors can be identified in this fashion.

The companies respond, "Well, that's better than nothing." On the other hand, it's a lot closer to nothing than it is to something; and further, given the nonspecificity of the match in the first place, it may actually be worse than nothing.

Scientifically, that is.

Emotionally or symbolically, it may be a lot better than nothing. And that's the beauty of this scam. The companies aren't scamming you. They are not giving you fraudulent information. They are giving you data, *real* data, and allowing you to scam yourself. And the result is that you, as an African American, feel good about having African roots—although you in fact have no more reliable knowledge of those roots from the DNA test than you had by simply looking in the mirror.[27]

A related venture addressed to Europeans involves identifying mitochondrial ancestors who lived about seventeen thousand years ago.[28] But its hard even to talk about ancestry in any biologically meaningful sense that far back. Mitochondrial DNA identifies one ancestor every generation (your mother's mother's mother's . . . mother); but the number of actual ancestors doubles as you go backward each generation, since everyone in the previous generation had two parents. If you figure four generations per century and 170 centuries, that's 680 generations. And the number of ancestors you had back then, in addition to the one mitochondrial ancestor, is a number with 204 zeroes after it—unimaginably large and effectively beyond the power of language to express.

That is also about 195 orders of magnitude greater than the number of people alive at that time, seventeen thousand years ago. Put simply, the great bulk of those ancestors were common ancestors, appearing multiple times in your own ancestry and of course in everyone else's as well. So you have one mitochondrial ancestor identifiable, who probably recurs a squijillion times in your own pedigree, and you have a schlemillion other ancestors also recurring a squijillion times. And so does everybody else.

When you go that far back, you are probably as closely related to Elvis of music, Akhenaten of monotheism, and General Tso of chicken as you are to your mitochondrial mama.

HEALTH AND MEDICINE

But surely there is a good reason to acknowledge race in health care. Don't you want black people and white people to have access to high-quality health care? Isn't that racial? And if it's racial and medical, isn't that biological? And if it's biological, then isn't race real? Give me more money.

This line of argument is particularly vile because it latches on to a real social issue—disparities in health care—and, whether naïvely or not, uses it to support propositions that are not quite humanitarian.

Certainly there are disparities in health care, and they are consequences principally of income, which is a product of social history, not a natural attribute like the air, the trees, and the grass. Further, ancestry is a risk factor for many health issues. Being Ashkenazi Jewish is a risk factor for Tay-Sachs disease, being Pennsylvania Amish is a risk factor for Ellis-van Creveld syndrome, being Greek is a risk factor for beta-thalassemia, being Afrikaans is a risk factor for porphyria variegata. Yet none of those populations is a race in any familiar modern sense of the term, so the question of ancestry as a risk factor in health has no bearing on the issue of the naturalness of racial categories.

Further, ancestry is only one factor in understanding health risks. Occupation is another: computer programmers are at higher risk for carpal tunnel syndrome, elementary school teachers for low-grade viral infections, miners for black lung disease, soldiers for being shot, and prostitutes for gonorrhea. Age is a risk factor for some things, and so is neighborhood.

The point is that knowing someone's ancestry is useful in understanding their health risks, but so are a lot of other things. Blacks and whites in the United States have some different health risks: whites are at greater risk for cystic fibrosis, and blacks for sickle cell anemia. Black American women are also at greater risk for having babies of low birth weight, and black American men for hypertension and associated cardiovascular problems. Yet African women immigrants are less afflicted by the low birth weight problem, and African men outside the United States are less afflicted with the hypertension problem—which makes it seem as though ancestry is a factor but not a directly genetic factor. The risk seems to come from growing up black in America.[29]

Knowing what we do about race, health disparities, and predictive risk factors, is it reasonable to seek different race-based interventions? We can begin by removing race from the question: does it seem reasonable to treat the same disease differently in different groups of people in the absence of other knowledge?

Probably not. All other things being equal, treatment for the same condition ought to be the same. But suppose there are genetic differences that affect the metabolism or efficacy of certain medicines? Then, obviously, they should be taken into account. But human genetic variation is primarily structured polymorphically and clinally. In other words, the hypothetical genetic difference is not likely to be present in, say, Asians but not Africans, but rather to be present in 42 percent of one and 63 percent of the other. The decision of whether to administer the medicine, then, has to be based on the determination of the patient's genotype, not their race. If there is one thing we know about human genetics, it is that race is a very poor surrogate for genotype.

On the other hand, genohype is a good surrogate for genotype. They only differ by one letter, after all.

Putting together two generic drugs that might be useful for the treatment of heart disease, a study conducted in the 1980s called V-HeFT (vasodilator heart failure trial) left the Food and Drug Administration unimpressed. The combination of drugs had been patented and was owned by a company called Medco, but when the FDA failed to approve the drug—now called BiDil—the company's stock took a steep plunge. Hoping to find a silver lining for this epidemiological cloud, Medco found that a subset of the patients responded a bit better than the others—the black patients. Medco sold the rights for BiDil to another company, NitroMed, which courted and received active support from the Black Congressional Caucus and the Association of Black Cardiologists. Framing another trial in the language of redressing disparities of health care in underserved populations, NitroMed arranged for this one to have a different regimen (so it was not strictly comparable to the first) and to include only black patients. In 2002, NitroMed received a patent for BiDil as an African American drug, extending by over a decade the rights they had when it was merely intended as a drug for all. The new study (A-HeFT, African American heart failure trial) found that BiDil was indeed effective—but *not* that it worked better than, or differently from, the way it did in other kinds of patients.

Regardless of the success or failure of BiDil—which failed to make NitroMed rich because the company grossly overpriced it—its significance lies in opening the door to a "racial pharmacogenomics," in which people are prescribed medications on the basis of their census category instead of their actual genetic makeup. A cynic might well conclude that this is not really about improving the quality of health care for the underserved at all but about opening up new racialized niche markets for pharmaceutical companies.[30]

The first step in this strategy would necessarily be to reify the racial groups as genetically bounded and homogeneous entities, which of course they aren't. But with so much potential profit at stake, perhaps it would be in the interests of big pharma to build an alliance with other groups interested in reifying races for other reasons. Thus, when Sally Satel published a *New York Times* op-ed titled "I Am a Racially Profiling Doctor," stumping for racialized pharmacogenomics, her affiliation with Charles Murray's *(The Bell Curve)* neoconservative think tank attracted some attention.[31] After all, redressing social injustices is not high on their agenda; denying those injustices is what they're primarily interested in.

Then, finally, with enough genohype you can even produce a new generation of biologists who actually think that they speak for science, and against political correctness, when they promote vulgarly racialized medicine. Sometimes this comes with an argument that scientific discussion of race is being stifled and that scientists are afraid to discuss it, much less acknowledge its verity, for fear of being censured by the Left. The speaker then becomes a lone objective voice for science and reason, unswayed by the social pressures and ideologies that are distorting everyone else's views.[32] The "suppression by the Left" argument isn't new, either; it was invoked by segregationists like Carleton Putnam, whose 1961 book *Race and Reason* didn't mention the Left but did run on about the conspiracy of communists, Jews, and anthropologists to stifle the obvious truths about race differences. A little quixotic, a little self-inflated, a little paranoid—and a lot anti-intellectual.

And just a bit evil, too.

Today those folks rarely look at themselves in the mirror and see a racist ignoramus, much less a shill for big pharma, staring back.

EVOLUTIONARY PSYCHOLOGY: IS IT EITHER OF THEM?

In the 1970s a newly named science burst onto the academic scene: human sociobiology, the application of evolutionary theory to the understanding of human behavior—as if nobody had ever tried that before. In fact it had been tried and had been shown to be facile, if not downright ridiculous, in every generation since Darwin.

The first generation of Darwinians, led by Herbert Spencer, saw the survival of the fittest as a good thing in nature and in society. They used Darwinism as a cudgel to rationalize exploiting and even extirpating nonwhite societies outside of Europe and poor people within their own society. The English paleontologist William J. Sollas put it this way in his 1911 book *Ancient Hunters:*

> Justice belongs to the strong, and has been meted out to each race according to its strength; each has received as much justice as it deserved. . . . It is not priority of occupation, but the power to utilize, which establishes a claim to the land. Hence it is a duty which every race owes to itself, and to the human family as well, to cultivate by every possible means its own strength . . . [lest it incur] a penalty which Natural Selection, the stern but beneficent tyrant of the organic world, will assuredly exact, and that speedily, to the full.

The second generation of Darwinians, led by the German Ernst Haeckel, saw not so much the parallel tracks of selection among organisms and among nations as the inexorable emergence of progress and order from chaos. Thus, they saw a single track leading from the lowliest amoeba up to the highest form of life and existence, the Prussian Nordic militarist state. I'm sure I don't have to tell you where that went.

The third generation of Darwinians, led by the American Charles Davenport (see chapter 3), conceptualized social history in terms of the distribution of hypothetical discrete Mendelian alleles, particularly the one for feeblemindedness.

The fourth generation of Darwinians, informed by the Nazi menace and led by such scholars as Theodosius Dobzhansky, Julian Huxley, and George Gaylord Simpson, coalesced Darwinism into the "Synthetic Theory" and acknowledged the separation of human history from the gene pool—as anthropologists had been arguing for decades.

Even so, in the 1960s a new wave of anti-anthropological Darwinism began to flourish, a view the previous generation had even named in deprecation "nothing-butism." To Julian Huxley, this involved "realizing that man is descended from a primitive ancestor, [and concluding] that he is only a developed monkey," a version of Darwinism that nevertheless produced science best sellers such as *The Territorial Imperative* and *The Naked Ape*.

Sociobiology emerged in the 1970s as an amalgam of all of these. Sometimes it exhorted readers to imagine a gene for altruism and how it might spread. Sometimes it exhorted readers to imagine the spread of units of culture, divorced from biology but analogous to genes. Sometimes it interpreted the behavior of baboons or chimpanzees as if it were easily confused for that of humans. Sometimes it ventured to explain all of history in terms of greater or lesser success in reproduction. And sometimes it invited readers to see non-Europeans as primitive actors in a Hobbesian "war of all against all" that the readers themselves have successfully transcended.

One way or another, Darwin's name was being dragged through the mud again—for this was presented as a "scientific" alternative to whatever fluffy nonsense the social scientists (and Synthetic Theorists) maintained. By the 1990s, however, human sociobiology had itself speciated. Veering off to the left was "human behavioral ecology," concerned with interpreting all aspects of human behavior in terms of their supposed adaptive functions. And veering off to the right was "evolutionary psychology," which took up the case for naturalizing the status quo, or explaining the way things *are* in terms of the way they *have to be*.

Evolutionary psychology effectively became the version of sociobiology that critics had dreaded from the outset. The central argument is that the gene pool has been shaped by selection; that the structure of the brain, like that of the foot, is a product of the genes; that the mind is a product of the brain's structure; and that ideas and behaviors are products of the mind. Looking for common patterns of thought and deed (and finding them more readily than other, more circumspect, students of human behavior), evolutionary psychologists localized them to mental "modules" that are themselves the hypothetical outcome of hypothetical selective forces upon the gene pool. To challenge any of these tenets—that there is a broad and readily discernible uniformity of mind that transcends its local variations, that all of its attributes are the direct products of natural selection, that the mind is modular in structure, and that it can be decomposed into individually evolving elements—is to invite the charge of being a creationist.[33]

The problem is that these scientists no more speak for Darwin or Darwinism than the segregationists or the eugenicists did when they tried to tar their own opponents with the brush of creationism.

Which brings us to the continuity between them. What the evolutionary psychologists have managed to do is legitimize an intellectual space in which to rationalize differences of gender under the banner of evolution. And although the theories and methodologies are different, the epistemology and rhetoric are remarkably continuous. Where the evolutionary psychologists talk about women and men, just substitute "blacks and whites" and you'll see what I mean.

Their bodies are different.
Their brains are different.
They behave differently.
The social differences are ubiquitous.
Sure there are exceptions, but look at the great differences in the averages.
Your common sense, or intuition, or folk knowledge was right after all.
Those liberal social scientists have been lying to you.
This is evolution, this is real science.

If you apply these sentences to race you produce the normative ideas of wealthy Americans in the 1910s and the shrill cries of frustrated segregationists in the 1960s. Today considerably fewer people think it, and when these ideas come to the surface they are usually either disguised (as in *The Bell Curve*) or recanted, in the knowledge that the position is not really scientifically defensible (as in James Watson's remarks).

The extraordinary accomplishment of evolutionary psychology, then, is to have opened up that *kind* of reasoning again—and not infrequently at that *level* of reasoning—as a legitimate scientific discourse when discussing men and women. Now, men could be from Mars and women from Venus—metaphorically speaking, for we know that there really is no life on either planet—without any hint of a naturalistic basis for those differences. Accountants might be from Saturn and choreographers from Mercury. PC users might be from Jupiter and Mac users from Neptune.

One of the most widely cited works in the area of evolutionary psychology purported to identify significant differences in what people say they are looking for in a mate—men in different parts of the world preferring young babes, and women preferring sugar daddies. A bit more sophisticated thinking showed that the preferences are strongest where women have the least access to resources. Far from being a global hard-wired response, it was much more likely a rational solution to a common problem.

Another widely cited work is almost embarrassing to describe. Shown silhouettes of women's figures, male college students in Texas overwhelmingly "prefer" women shaped like Marilyn Monroe. That is to say, dividing the last two measurements of the 36–24–36 starlet's figure, they settle on a waist-to-hip ratio of 0.67. Male college students in many places express a similar preference. The "evolutionary" explanation is that this reflects an innate drive toward the perfectly evolved woman, deviations from which men accept only with some reluctance, as it were.[34]

The equally evolutionary, but less idiotic, interpretation is that it instead represents the diffusion of contemporary American media tastes and values to the rest of the world—in a word, globalization. There is, obviously, a crucial experiment one can perform: find some very remote people and discern their tastes. Surely enough, the short and stocky Matsigenka men in highland Peru prefer their women short and stocky; and the tall, thin Hadza men in East Africa prefer their women tall and thin.[35]

There is a crucial cautionary tale associated with this research, however. These kinds of critical experiments are becoming more and more difficult to perform, as the economic and social forces entangled in American popular culture reach even the most remote peoples on earth. When everyone has been exposed to the same cultural information and values, it will be impossible to distinguish those broad uniformities that are the result of being human from those regularities that are the result of living in an increasingly homogeneous society.

The crude deduction of innateness from the observation of similarity or difference is what evolutionary psychology regressively provides. In the mid-twentieth century, paralleling the separation of naturalistic "race" from culturalistic "ethnicity" (a separation now seen as a bit too facile, for races have highly constructed aspects, and ethnicities have some naturalistic ones,[36] but which analytically was very valuable for its time), students of human behavior began to separate naturalistic "sex" from culturalistic "gender." Sure, drawn from the same population, women average 25 percent smaller in body mass than men (which parallels sexual dimorphism in the great apes) and have a higher proportion of their body taken up with subcutaneous fat (which actually doesn't parallel the great apes). But those facts of nature ought to be irrelevant to the questions of enfranchisement and employment. Margaret Mead helped to document the diversity of roles women assumed in cultures outside the Western mainstream and forced readers to imagine a society almost exactly like their own but in which women's life options were not quite so constrained.[37]

Perhaps it is just a coincidence that evolutionary psychology began to emerge just as the conservative backlash against the Equal Rights Amendment peaked in the early 1980s. But does it really matter what kinds of brains women have? Nobody really thinks that brain structure is an independent variable in human behavioral development anymore, except in pathological cases, do they? And yet a notorious comment in 2005 by Harvard's then-president Larry Summers—about a larger proportion of men at the high end of the sciences probably being naturally endowed for success in science—made it clear that the issue is very much alive.

Summers, however, was actually only reiterating what he had heard from evolutionary psychology.[38] And the most important thing he heard wasn't about evolution. It was about the low percentage of women being promoted in the sciences at Harvard *not being Harvard's fault*—it was *women's* fault. In other words, like previous versions of hyper-Darwinism, evolutionary psychology was just affording a biological rationalization for the status quo. In this way a perceived problem can be denied rather than assessed—and without even a working knowledge either of evolution or of psychology!

The thoughts about human heredity expressed by the president of Harvard were decidedly premodern and largely independent of the world of Gregor Mendel, Theodosius Dobzhansky, Victor McKusick, or even James Watson.[39] It's not about transcribing, interacting, and reassorting bits of nucleic acid. It's about: Look at 'em! They're different! Everybody has what they deserve! Here's the money you asked for!

The big irony is that Summers was ultimately replaced in his job by a woman.

DARWIN AS CULTURAL ICON

The fact that Darwin can be so easily co-opted for causes like rationalizing inequality should give us pause. What does naturalizing social injustice have to do with the propositions that the patterns in the diversity of life have a genealogical basis or that adaptation is historically produced and not an endowed state? Those are the central issues of *On the Origin of Species*—that is to say, of Darwinism.

Obviously we aren't talking about Darwin's *Descent of Man,* which is a fine book but certainly encodes the premodern social values of Victorian England throughout its text. Nor are we talking about Darwin's *Variation of Plants and Animals under Domestication,* published in 1868. That is the one in which Darwin proposed his famously wrong theory of heredity known as pangenesis, which had the body parts secreting little buds, or gemmules, that traveled through the body's fluids and coalesced in the reproductive organs. Not only was it wrong, but it was old hat. He shared the theory with Thomas Huxley, who told him that Buffon—the French naturalist of the previous century—had been there first. Darwin wrote him back, "I have read Buffon:—whole pages are laughably like mine. It is surprising how candid it makes one to see one's views in another man's words."[40]

But there is an odd quirk of history associated with that incorrect theory of Darwin's. Around the beginning of the twentieth century, the early Dutch geneticist Hugo de Vries was grappling for a neologism to apply to hypothetical elements of heredity. Darwin's shadow loomed so large by that time that de Vries thought of Darwin's pangenesis and named the units of heredity in his honor: *pangenes.* In a 1909 textbook, the Danish geneticist Wilhelm Johannsen honored Darwin by adopting de Vries's term and dropping the initial syllable. Thus does Darwin come to be the father of the *gene*—although by a different route than he came to be the father of evolution. Darwin's work in evolution was paradigm defining, but his work in heredity was forgettable. And yet he is commemmorated in both fields.

Darwin is a cultural icon.

There's Darwin in literature. There's Darwin in archaeology. There's Darwin in medicine—ironically, a field he dropped out of as a student. A Darwinian medicine certainly sounds reasonable, on the face of it.[41] But a Darwinian medicine could actually encompass a wide range of ideas, from infanticide, to identifying antibiotic resistance in bacterial pathogens, to the racial pharmacogenomics ushered in by BiDil.

Darwin is benefiting from a spillover effect of being the eponymous leader of a scientific movement. There are a lot of Darwinisms out there, but his name is the one attached to them all. With so much being brandished in Darwin's name, we need to be vigilant about keeping it unsullied, as Clarence Darrow realized at the time of the Scopes trial. At its most basic, *evolution* is a complex homonym, referring to four entirely different things, which occur to different objects, at different rates, and via different modes. First, cosmology, as in "the evolution of the solar system." Second, ontogeny, as in the "evolution" of a fetus into a codger—and although this goal-oriented sense sounds most foreign to us, this was actually the primary use of the word in Darwin's time. Third, the diversification of species and their consequent adaptation through natural selection—the narrowest and most appropriate sense. And finally, the emergence of cultural diversity over the much shallower time frame of social history—as in the "the evolution of baseball."

So, assuming that it is good—that is to say, modern, scientific, wise, biological—to be a Darwinian, then what form would a Darwinian study of human behavior necessarily take? Would it involve racism? Would it involve believing that we are living in the best of all possible worlds, shaped only by the competitive replication of genetic elements? Would it involve modeling the competition of "memes" instead of genes? Would it involve casting a blind eye to injustice and declaring it to be the law of nature? Would it involve killing babies? Would it involve interpreting the history of the world as if everyone were actually maximizing their breeding? Would it involve trying to explain how everything—including smoking, homosexuality, altruism, and divorce—is adaptive? Would it involve believing that genetics is at the root of all interesting questions? Would it involve believing that the minor differences in thought and deed *within* a group of people, where behavioral genetics may be a contributing factor, can unproblematically be extrapolated to explain the differences in thought and deed *between* groups of people?[42]

Or would a Darwinian anthropology still focus on questions of power, gender, and difference—and continue to interpret them in the context of politics, economics, and meaning—and simply be *compatible* with the proposition that humans evolved from apes, in the same fashion that it is compatible with the sun being the center of the solar system, or with masses attracting one another in proportion to the inverse square of the distance separating them?

If you oppose so-called Darwinian approaches to human behavior, are you a creationist? The advocates of such approaches would like to believe so and have had occasional successes at making that association. On the one hand, Thomas Huxley had unimpeachable Darwinian credentials when he debated Herbert Spencer on the merits of legislatively curbing the ruthlessness of evolutionary "selection" in modern society. On the other hand, at about the same time, the great German biologist Rudolf Virchow was forced to take sides by Ernst Haeckel on the evolutionary superiority of one kind of person and state over all others. Virchow chose to reject all evidence for human evolution on the suspicion that any such evidence might be brandished by the Haeckelians on behalf of their odious political views. (And he was right.)[43] Charles Davenport and Madison Grant would give a similar choice in New York a generation later to Virchow's former protégé, Franz Boas: to be against us is to be against Darwin.

But let's back up. What would someone principally interested in history, cultural diversity, human agency, and social justice stand to gain from Darwinism? Common biological descent and adaptive divergence aren't really the subjects; Darwinism is a red herring here. Its force stems from the recognized power of the scientific revolution Darwin helped to catalyze, but his application to human affairs is only metaphorical. What, for example, might a Copernican anthropology be like—or, more properly, would an *anti*-Copernican anthropology be any different? What about an anti-Newtonian anthropology? It seems to me that the practice of understanding human diversity is not significantly affected by your view of the solar system or of gravity—although if you tried real hard you might be able to connect them.

Actually, the study of human behavioral diversity is in a good position to deflect the question "Why can't you be more Darwinian?" After all, being founded on cultural relativism, there is a case to be made that anthropology is past Darwin; it is already Einsteinian.

The evolutionary geneticist Theodosius Dobzhansky pointed out decades ago that identity and equality reside in different spheres. Only monozygous twins are genetically identical, but the state decides that its citizens are equal under the law, regardless of their biological diversity. The biological fact of difference is unrelated to the social fact of inequality. They can be related in the sense that all societies incorporate different people of various kinds and assign meanings to those differences. But equality is about cultural processes, deciding which differences are important and which are not. Arguing about the brains and genes of different groups of people is thus largely tangential to the formation and maintenance of a just society.

We Darwinians are the ones who have to clean house, to deny the label of credible and authoritative science to the metaphysical and metaphorical biology employed in evolutionary psychology. If the role of science, in this case evolution, is to provide a spurious foundation for anti-democratic discourses in the modern world, then does it not follow that we would be better off without it?

The key lies in conceptualizing humans as simultaneously continuous with and divergent from the "other" apes. Sure, we have short, stiff spinal columns, fused caudal vertebrae, no tail, a posteriorly positioned scapula, and a rotating shoulder, as they do. On the other hand, our brains are three times the size of theirs, and we're walking and talking and they're not. Those are the twin pillars of Darwinism: descent and divergence. If we focus, as Darwin did, on "the origin of species," then the latter should be the more important of the two. The origin of species is divergence. But the call to Darwinize anthropology is generally a call to focus on descent at the *expense* of divergence, which is in turn the classic reductive agenda of "nothing-butism." In a critical sense, it is a theory of human evolution that begins by assuming we never really became human.

Consider the distal hindlimb of chimpanzee and human—one adapted for grasping, the other for bearing weight. They look rather alike; they are made of pretty much the same parts in pretty much the same relations. But if you are interested in the human foot—how it works, what it means, even where it came from—there is very little that studying a chimp foot can tell you, except by way of contrast, that you cannot learn better from studying *human* feet. A chimp foot can be trained to bear weight to some extent, and a human foot can be trained to grasp to some extent. But what is interesting about the two feet, from the standpoint of evolution, is how they differ.

That is also why, in spite of having corresponding forelimb parts in similar relations to those of a sparrow, you still cannot get off the ground by flapping, while the sparrow can.

Returning to the ape, then, if locomotion is so different between us, and the structures have been reworked so as to alter the basic function, then what about cogitation? Is it possible that the threefold growth in size, extensive cortical convolutions, neurological reorganizations facilitating speech—that all those things make human thought and behavior different from—not bigger than, not more complex than, not a variant of, just *different* from—chimp behavior?[44]

I believe they do. The value of chimp feet for understanding human feet lies in their contrast, not in their sameness. Likewise the value of chimp brains and behaviors lies in their contrast to humans. This is not about piety or humanism; it is about epistemology, methodology, and rhetoric. The feet are adapted to different purposes, and the brains are adapted to different purposes. Labeling things that look different and do different things as "the same"—because their parts roughly correspond and six or seven million years ago their *ancestors* were the same—is not only weird, it is perverse. It is *anti*-evolutionary—especially if that labeling is used to encode an argument for the natural inequality of large groups of people, in Darwin's name.

Consequently, I don't place much stock in this primatologist's evaluation: "In their emotions, cognition, linguistic ability, homicidal brutality and erotic sexuality, the apes and we are far more alike than we are different."[45] I'd sure like to know what this likeness in, for example, "erotic sexuality" means, if a chimpanzee male is stimulated by purple estrus swellings and copulates for fifteen seconds. Is it possible that we have produced a generation of hyper-Darwinized primatologists who have come to know more about apes than they do about people?

The origin of species, said Darwin, lies not in theology but in adaptive divergence, and it is exactly that divergence that interests us as post-Darwinian biologists and anthropologists. In other words, it is *evolutionary* to acknowledge the difference of humans, and an evolutionary theory that fails to come to grips with that is not going to be of much use as an analysis of behavior, or as a representation of nature.

For if one person accepts the evolutionary divergence of human and ape, and another denies it, then who is really the creationist?

TEN Nature/Culture

I don't want to be a member of any club where scientific racism is welcome. Scientific racism has no place in science. Like scientific creationism, it is ideology disguised as science, to give it legitimacy. If it fools some people who have scientific credentials, that is a problem for the credentialing process and for the scientific community. An integral, if often overlooked, aspect of science is the ability to tell when scientists say things that are false, or evil, or both. Scientists themselves sometimes cannot even tell—because they are no better than anyone else (and possibly worse) at telling good from evil and are better at telling true from false generally only within their specific and restricted domains of expertise.

The rigorous study of nature, known first as the New Philosophy and then as science (see chapter 2), began, as significant cultural endeavors do, with the construction of boundaries.[1] Science effectively began with the erection of a conceptual wall separating what seventeenth-century European scholars considered the domain of God from the domain of His creation—that is to say, between supernature and nature. The sources of knowledge in each domain would be different: the former would be the domain of miracle and revelation; the latter would be the domain of generalization and experimentation. On one side, spirit; on the other, math.

Citizen-scholars could certainly be interested in both kinds of knowledge, but they would have to be compartmentalized and kept separate from one another. By the turn of the eighteenth century, Isaac Newton could separate the two so effectively that the recovery of his copious theological writings would actually come as something of a surprise. As we learn more of the diseases that germs cause, and how to treat them, our knowledge of the diseases caused by evil spirits has remained constant and no concomitant expansion in their treatment has taken place. The diseases caused by germs are understandable scientifically; the ones caused by evil spirits are not amenable to science and its methods. A cognitive wall has thus arisen to divide the capricious forces of the spiritual realm from the regular, and knowable, forces of the physical world.

From the other side, religion gives coherence to life in a fashion that science hardly attempts. Science is, after all, just a set of newfangled methods for trying to find out what's happening in the universe. It seeks accuracy, not coherence. It doesn't strive to impart meaning to life, or to discover meaning within it. That is for other kinds of study and knowledge.

The wall is slightly porous, however. Students of the supernatural can find meaning in the natural. Nature is amenable to nonscientific study, after all. Science itself is amenable to theological discourses. Natural theology (whose most famous expositor was William Paley and whose latest incarnation is intelligent-design theory), the science Darwin learned in college, assumes that God left His imprint on the world when He created it and devotes itself to discovering that imprint. But the assumption that the study of nature reveals aspects of supernature, by a process of easy translation, itself violates the formula that has been so successful for science over the past few centuries. When the move comes from religion (that is to say, when theologians apply what they know of supernature to the natural world), the result is generally boring but not nonsensical, because the boundary between the two domains was erected to circumscribe science and is not necessarily respected by nonscience. Religious ideas were there first, and many ancient peoples understood and manipulated nature, and made religious sense of it, before science came into existence. Science is constituted by the recognition of a boundary between nature and supernature, and that boundary does not necessarily exist outside of the minds of scientists. In the minds of nonscientists, it may seem quite reasonable to understand nature in terms of spirit.

The problem comes when scientists themselves fail to respect the boundary between nature and supernature and begin to apply the methods to the spirit world that work so well for understanding and manipulating the natural world. But when modern science tries to move through the boundary into religion, the move is inconsistent with the development of science itself and is thus nonsensical. The spirit world is impervious to the methods of science; we instead study scientifically what is amenable to the methods of science. In other words, the boundary is selectively permeable.

The most obvious self-interests of science in opposing or delegitimizing religion were clear in the late nineteenth century, when the doctrine of cultural progress predominated (see chapter 4). If you could see history as progress, and progress was driven by science, then it stood to reason that in the context of European history magic or superstition had yielded to religion several thousands of years ago, and religion was in the process of yielding to science now. Not surprisingly, prominent scientists such as the sociologist August Comte, the biologist Ernst Haeckel, and the geneticist Francis Galton all saw themselves as founders of something that would replace religion as we know it. Religion was only for people who couldn't handle science. These three, of course, would be sort-of popes. All three men are regarded as quaintly premodern founders by their respective disciplines today.

Stephen Jay Gould categorized the domains of religion and science as "non-overlapping magisteria"—but they can overlap, and when they do they play by different rules. In the case of the creationists, one side wishes to save your soul from eternal damnation; the other wishes to convince you that you came from a monkey. It stands to reason that they would perceive the issues, evidence, and rules of engagement differently.

THE OTHER WALL

Science came to be shielded on the other side, the human side, as well. While God may have created the universe, its laws and properties are the domain of science and can be studied in the same fashion whether or not God created them, and whether or not they have a reason for existing.

But what of the things God didn't make? Like standards of beauty, the pyramids, national boundaries, traffic laws, novels, your home, your job, your clothes, and all of your possessions? Those are the products of human history, human thought, and human labor. We will call them *humanities* or *culture* and lay off them as well; the rules of science don't apply to them either. Like supernature, this domain comprises another kind of knowledge.

This field had its intellectual maturation with the development of cultural relativism, after World War I. Progress could no longer be taken for granted, and science did not necessarily make the world a better place; rather, it broadened the potential for human suffering. So, by the later 1920s, the progress associated with a pseudo-Darwinian view of history and society could no longer be sustained. Just as things as natural and universal as mass and time may be dependent upon the perspective of the observer (if you happen to be moving close to the speed of light), so too can blurry cultural progress be resolved into a set of assets and liabilities, arbitrary conventions, and responses to particular circumstances by the mere act of positioning oneself outside the culture in question. And while technology appears to be progressive, from a broader (especially a "post-9/11") perspective, it is clearly a mixed blessing

So where does that leave the positivist assertion that religion has been superseded and is only for the young and the stupid? It leaves it in need of some justification, as with any other ethnocentric assumption. After all, follow a group of scientists at a convention in Las Vegas, Foxwoods, or Sun City and you will see that ideas of taboo, magic, animism, and superstition are very much with us, even among very smart people. They just separate their science from that stuff.

Once again, however, the boundary is selectively permeable. While scientists would like as much freedom as possible to pursue their studies of the universe, it has been appreciated since World War II that science must be subject to the constraints of decency—that is to say, science is not free of the moral realm, and scientists will be held accountable for work that is deemed offensive or evil. But since good and evil are defined by the time and circumstance, it stands to reason that there is no way to learn about them in the natural world. That information must come from the other side of the wall.

Bioethics is the most notable area in which ideas from outside of science come into science, as people trained to think about right and wrong try to negotiate that knowledge against the study of nature. Here, once again, the understanding is that in order to exist among the social creations of any human collectivity, science must be subject to its rules. Dishonest or evil science is not acceptable, but identifying it is predicated on the nonscientific knowledge of what is constituted by dishonesty and evil.

From the other side, though, scientists talking about law, morality, history, or aesthetics usually run a gamut from crassly self-interested ("ethics, schmethics"—chapter 8) to embarrassingly ethnocentric, still making cultural comparisons in terms of progress, as was standard practice a century ago. On the other hand, as C. P. Snow implied (chapter 1), they can't be blamed, for it's not their domain of expertise. The problem is the appropriation of cultural authority, as the historian Jacques Barzun observed:

> Where and what am I, whither bound and for what ends? These questions that man keeps asking, all agree that science cannot answer. But the confirming cliché—Science tells How not Why—is falsified in reality by the appearance of answer-giving which science has been guilty of for over a century. And when it has not so transgressed it has issued prohibitions against answers given by others.[2]

Most bizarre of all is to see this archaic positivistic thought at work within the scientific community today, as an aggressive anti-religion contingent takes on a two-front "science war." First, they oppose the creationists, which seems reasonable; but second, they adopt an obsolete prerelativistic approach to human history and culture in doing so and rail against the cultural relativists on their other front. Ironically, then, the contemporary evangelical scientific atheists are actually worse than the creationists, for the creationists don't operate within the framework of the normative practice and cognitive apparatus of science and are consequently at best on its fringes; but the evangelical scientific atheists misrepresent science even while claiming to speak for it. Creationists can't make science look bad; only scientists can do that.

If we have made intellectual progress in human science, it lies in appreciating the diversity and creativity in human thought to a much greater extent than earlier generations did. So for Darwinians to argue that the ethnocentric progressivist human science of a century or more ago was somehow better than what we have today is simply a replication of the anti-intellectualism they presumably abhor.

I sure hope the creationists aren't paying too close attention.

SCIENCE AND HUMAN DIFFERENCE

Significant changes in brain size and shape accompanied technological—that is to say, cultural—change over the evolutionary history of our species. Early members of the genus *Homo,* about two million years ago, created sharp edges on stones by banging them together, and they did this with a brain about the size of a gorilla's (in a considerably smaller body). Their descendants a million years ago, with brains half again as large, built on those techniques and developed standardized toolkits dominated by the "Acheulian bifacial handaxe"—a flat stone worked on either side and chipped carefully to a triangular shape This was a tool so impressive that to this day we are still not certain exactly what they did with it; but whatever it was, they did it well, since those tools were made for hundreds of thousands of years. Their descendants, the Neandertals, improved on the Acheulian toolkit and did so with brains the size of our own, although flatter and longer.

Our own ancestors of one hundred thousand years ago, with smaller faces and rounder heads, began with much the same stone-chipping capabilities but also decided to pierce holes in shells, decorate themselves and their surroundings, and expand the range of raw materials they worked on. And in no time at all, paleontologically speaking, they were painting on rocks and caves, growing their own food, writing, blowing things up, and surfing the Internet.

All of that last bit, however, developed without detectable difference in cranial structure. The rise from flint knapper to porn surfer has occurred mentally, but not cranially. In other words, cultural/historical evolution has largely replaced discernible biological change as the principal source of diversity and adaptation in the human lineage. It involves linguistic, rather than genetic, transmission, and involves the production of thoughts and behaviors that are locally specific. This has two important consequences for the study of human evolution. First, our coping skills—the information about who we are and how to live—are acquired throughout the course of our lives; they differ from place to place, time to time, and situation to situation. Second, there is a significant discordance between biological and behavioral variation in our species. The major features of human behavioral diversity are located at the boundaries of populations, differentiating human groups from one another, while the major features of human genetic or biological diversity are located within the populations themselves.

Whether deliberately or accidentally, the latter consequence is commonly misunderstood and misrepresented. It is foolhardy to deny that genetic variation affects human behavior. The question is, how significant is it? The great discovery of human sciences in the twentieth century was that there is pattern or structure to the human spectrum of behavioral variation. Appropriate behavior is necessary for survival, but it is defined locally and must be acquired; the very ability to communicate is likewise species-wide yet locally defined. At the very least, a person is a part of a linguistic community, and to the extent that one can participate in multiple communities it requires a degree of multilingualism. But it is these sets of rules and practices that principally distinguish humans as group members from one another. This local specificity is what gives human social groups their cohesion and identity and has come to be known as *culture* (see below). In the great scope of things, it is the realm in which the bulk of variation in human behavior lies.

There does not seem to be any significant relationship between genetic diversity and the ability to speak French, Korean, or Apache; what one speaks is due entirely to the circumstances of one's life. While the need to eat is species-wide, what is considered appropriate to eat or even deemed palatable varies in a similar fashion, very extensively and independently of the gene pool. Thus, while Leviticus 11:20 makes it clear that crickets are not only edible but kosher, you would be hard-pressed to find a Jew, observant or otherwise, likely to eat one today. The force of western European cultural history has rendered all insects meaningful as "nonfood" in spite of being edible and eaten in other places and times, including by one's own ancestors. There consequently doesn't seem to be much of a relationship between behavioral variation and genetic variation, for the behavioral variation seems to be based principally upon learned information.

To look at it a different way, the same brain hormone receptor allele in a Harvard professor and a Venezuelan Yanomamo would not make their lives any more similar. They might get through the day a little more happily, or more introspectively, or more addictively, than other Harvard faculty members and other Yanomamo, but no significant aspect of their lives would thereby converge—for what makes their lives so different lies not in their brain genes but in their social histories and circumstances. Thus, to the extent that the possibility exists of studying the effects of genetic variation upon human behavior, it can be useful only for studying those features of human behavior that vary as genetic differences tend to—within populations. That is to say, it would affect only the most minor features of the diversity of human thoughts and behaviors, and principally pathologies.

The interests of the behavioral geneticist and the racist differ, but they can converge if the discordant patterns of human behavioral and genetic variation are ignored. Then, genetic variation for behavior or even for intelligence, which may be useful in explaining within-group differences (such as those between normal children and those with Down syndrome, which is found everywhere), can casually be extrapolated to explain differences *between groups* in thought and deed. Such an extrapolation is entirely illegitimate, yet every generation seems to be confronted with it, and often in the name of science.

Why is this confusion so tenacious? In part, at least, it is so because of the convergence of interests between genetics and conservative social politics. Once you know that, for example, Koreans fare worse than Japanese on IQ tests in Japan but not in the United States, you realize that the social status of group membership counts for something, that test score differences are not immune to the effects of cultural difference, and that simply framing discussions of intelligence as if it were "genetic nature vs. environmental nurture" serves to erase the cultural context of the data, which may well be the source of the group-level difference detected in the first place.

If your primary interest, however, lies not with science but with keeping social divisions secure, then you may well have a convergence with the geneticists, whose interests lie in establishing the importance of their science. Genohype thus becomes a win-win situation, as both the political theorist Charles Murray and the molecular geneticist James Watson recognized. Talking about the genes in the context of social issues keeps the focus off institutionalized inequality, making it nobody's fault but Mendel's and Darwin's.

With such conflicted interests, we have to examine any claim about human behavioral genetic differences very carefully. In 1937, the great Harvard physical anthropologist Earnest Hooton, who was by no means a postmodernist but was at great pains to distinguish the bad Nazi study of human variation from the good American study of human variation, put it this way:

> There is . . . a rapidly growing aspect of physical anthropology which is nothing less than a malignancy. Unless it is excised, it will destroy the science. I refer to the perversion of racial studies and of the investigation of human heredity to political uses and to class advantage. . . . the output of physical anthropology may become so suspect that it is impossible to accept the results of research without looking behind them for a political motive.[3]

Hooton was right. The cultural context in which this science takes place makes it so. You can't separate the science from the culture; the culture makes the science possible. It validates science as a way of knowing reliably about the world, and it validates this science in particular by directing resources (intellectual, institutional, financial) toward it. Can it be surprising that the first generation of scientific studies of human diversity in the United States—the so-called American School of physical anthropology—came with an explicitly pro-slavery agenda? Or that the next generation produced the social Darwinists, and the generation after that produced the eugenicists? Then, in the three generations after World War II, the segregationists, sociobiologists, and evolutionary psychologists in succession? All have claimed to speak on behalf of science, indeed on behalf of genetics, in their different ways—although with the exception of the eugenics movement, rather few practicing geneticists actually joined these scientific camps. These have consequently tended to be discourses about genetics by nongeneticists.

If that sounds contradictory, perhaps it is worth pondering. Geneticists generally don't think of genes as "selfish," for example. They tend not to think that IQ is innate, nor that specific mental properties can be causally associated with specific alleles. They tend to be politically liberal, like other academics. So why this convergence with anti-democratic political ideologies? That is precisely the question that geneticists needed to ask in the 1930s in the first person: "What is it about me that the Nazis like so much?"

The answer is that human genetics defines the scientific discourse that overlaps the folk ideology of what is innate and natural, which in turn helps to delimit what is just and permissible. The argument of *The Bell Curve* (1994) was fundamentally that average differences in IQ between blacks and whites were the result of natural process, not social injustice. Thus programs like Project Headstart would be solutions to problems that didn't really exist.

The Bell Curve is a book that made claims about genetic differences in humans, by a political theorist and a psychologist. It self-consciously cited work that was outside the scientific mainstream—published in racist forums, sponsored by a notorious philanthropy that has historically had a soft spot for eugenics, segregation, and scientific racism. *The Bell Curve* also presented the work of Philippe Rushton—a Canadian psychologist who argues for the IQ of native Africans being indicative of marginal mental retardation, who uses head size as an indication of intelligence and penis size as an indication of sexuality and reproductive rate[4]—and defensively told readers in an appendix that his work demands to be taken seriously.

Of course it doesn't. The point is that all scientific work isn't equal.

Then a writer for a popular e-zine, Slate, connected the dots—putting together Bruce Lahn's work (chapter 6), Philippe Rushton's work, James Watson's claim about the genetic stupidity of Africans (from which Watson himself quickly tried to retreat, to the chagrin of conservative pundits), human nature, genetic determinism, and racial essentialism—and, of course, dragged the good name of Darwin through the mud in all this. A bit later, the medical literature started to see opinion pieces on how Watson might really have been right after all, on how he had been stifled by that old left-wing conspiracy, resurrected from the days of the segregationists.[5] Watson, it seems, had inadvertently shined a light on the old/new scientific racism.

But this is simply the result of successful propaganda in the service of legitimizing inequalities and is more closely related to the work of scientists like Charles Davenport than Charles Darwin. The politics—the culture—is ubiquitous, and consequently the science is not, and can never be, free of it. In this case, then, science education is at least as much about making sense of the relevant social facts that help produce the natural facts as it is about the "naturalness" of the facts themselves. Even the assumption that the facts about human diversity can readily be distilled into contextless "natural" facts and separated from the cultural environment that coproduces them, in any but the crudest of fashions, is itself gratuitous—as Earnest Hooton was realizing back in 1937.

NATURE/CULTURE

There is probably no literary genre worse than theologians writing about science, except scientists writing about theology. The boundary between spirit and matter is there for science, not for the rest of us. And the fact that scientists have done so well with nature does not necessarily mean they have anything interesting at all to say about supernature.

Or consider it from the other side: whom do you want by your side when the IRS comes to audit you—your friend the accountant or your friend the X-ray crystallographer? You want someone who knows the tax code, which is a testament neither to the mind of God nor to His works on Earth. It is the product of human beings and consequently is not of nature but of something else. Let's call it, as they started to do in the nineteenth century, culture.

Crucial ideas are notoriously difficult to define rigorously, for they are crucial because they make a novel distinction—by giving something a name that was previously unnamed. Once scholars find the distinction useful, they approach it in different ways and begin disagreeing on what its most salient features are, and what to do about boundary cases. In biology, *species* and *gene* come readily to mind; the more you know about them, the fuzzier they become. That doesn't mean they don't exist, just that they are often defined more heuristically than rigorously, and any particular definition reflects a particular frame of reference.

The same is true of culture. The distinction it marks (from nature) is significant, for it separates the optic scanner from the eyeball, the Empire State Building from Niagara Falls, the senior prom from the courtship dance of the honeybee, and the Constitution from the Pythagorean theorem—that is to say, the artifact from the fact. On the other hand, culture eludes an easy definition: cultures vary or vanish according to the perceptions of the observer, they can be subdivided endlessly, a single person may partake of multiple ones simultaneously, and they seem to be unbounded by time or space.

One problem is clear: we use the term *culture* in two very general ways. One is as an antithesis of *nature* and a synonym of *civilization* (the original, nineteenth-century sense, reflecting a balance between the two, with Europeans very civilized and aborigines very natural). The other is as the effect of history, upbringing, and circumstances upon one's particular behavioral patterns—the ubiquitous local prejudices that suffuse word, thought, and deed (the twentieth-century, relativistic sense, in which everyone is equally and completely cultural).

It is this newer usage of *culture* that is often grafted onto an earlier, pre-Darwinian concept of human nature and seen as a model for the scientific analysis of the human condition. Indeed, it is often even presented as virtually part of a Darwinian "Pledge of Allegiance," such that to question it is to risk being called a creationist.

Mel Konner observes that "some still deny that human nature exists.

It is difficult to understand what this denial means."[6] To me, it means that the search for some ubiquitous, innate, immutable, and uncultural features of human thought and life is a fool's errand. Generally it reveals information that is trivial ("It is human nature to be grammatical.") or varying degrees of inane ("It is human nature to want to pair up with a young hottie or a sugar daddy." "It is human nature to get divorced after seven years of marriage." "It is human nature to rape.").[7] This is so because inherent in the concept of human nature is that it can be divided from culture; that culture needs to be scraped off, like the icing on a cake, to reveal the human nature below. But if culture is not so much like the icing on a cake as like the eggs in a cake, then a program that involves isolating it from human nature is not likely to be of much use.

That's the problem.

I am dwelling on this point for a reason: Culture is inseparable from being human. And yet many scientists still get it backward, believing that they can separate a deep human "nature" from a superficial human "culture," and often in the name of science.[8] But culture is also the product of millions of years of coevolution with the human species and cannot be so casually separated from it. We can argue about what chimps are capable of understanding and of making, but the fact remains that this is *our* evolutionary adaptation, not theirs. You can survive as a noncultural chimpanzee, but not as a noncultural human. We have been coevolving with art for over one hundred thousand years, coevolving with tools for over two and a half million years, and coevolving with language for probably as long. Culture is not so easily pushed aside.

Moreover, culture is there as both an ultimate and a proximate cause of the human condition. As an ultimate cause, culture provides the social, cognitive, and technological environments to which our gene pools have had to adapt over the past few million years. The best-known case of genetic selection operating on the human gene pool is that of resistance to malaria—which itself became a major health stressor only *after* the mosquitoes began to visit the standing water produced by early farming, that is to say, within the past ten thousand years. And culture is a proximate cause as well, in that human phenotypes are coproduced by genes and environment, and the human environment is invariably cultural. We still do not know why Sicilians tend to have narrower, longer heads than Russian Jews, or why the head shapes of both groups tend to converge in their American descendants, nearly a century after Franz Boas first took the measurements. We do know that strapping your child to a cradleboard will make the child's head flat at the rear, or, more broadly, that the human body is developmentally plastic and highly responsive to the conditions of growth—and that those conditions are cultural.[9]

So where does that leave the program of discovering a cultureless human nature? Not particularly well off, obviously. Like many things metaphysical, it's a lot easier to assert than to demonstrate. Usually it involves (1) imagining that everyone is, and has always been, just like you; (2) mistaking broad generalizations of what humans do as the products of human nature rather than as effects of globalization or simply reasonable solutions to common problems; (3) ignoring exceptions to the generalization, which renders those exceptions either unnatural or not human; and (4) invoking the generalization to rationalize familiar social or political hierarchies.

This is largely the gamut of premodern anthropological errors: ethnocentrism, nonrigorous method, dehumanizing subjects, and perpetuating injustice.

A more reasonably Darwinian mode of thought might involve recognizing that there are few things that all normal people do. At the very least, we can see walking and talking as biological adaptations. Yet both are actively learned in childhood, and in locally specific manifestations. We are programmed to learn language, but we learn the languages we are exposed to. We are programmed to learn to walk, but body movement is also highly culturally specific.

Further, it is just as human to sit down and shut up. In other words, the knowledge of when and how to perform these bits of so-called human nature is also locally specific, and *not* doing them may be as much a part of "human nature" as doing them. If it is human nature to smile, it can nevertheless get you killed if you do it in contextually inappropriate circumstances. It is only human nature to walk, talk, and smile *when you're supposed to.* So what really is the value in talking about human nature as if it ought to, or could, be analyzable or meaningfully understood independently of culture?

Consequently, human nature is not something to be acknowledged or denied but rather something to be un-thought. The concept of human nature has nothing to do with Darwinism—it was there long before Darwin—and is anti-evolutionary in imagining a static gene pool unaffected by millions of years of human coevolution with sets of rules. That is why other, more powerful, metaphors are commonly invoked by the scholars who don't find the concept of human nature to be particularly realistic or useful in discussing the human condition. One is to talk about a subject called nature/culture; another is to conceive of an active process by which social and natural forms coproduce each other; and yet another is to emphasize the human phenotype itself as being a "cyborg" object—invariably both natural and cultural.[10]

One thing is clear, though: Darwinism doesn't stand or fall on "human nature," and invoking it in such a context has far more rhetorical than scientific value.

That thought, in turn, helps provide the most important insight about race. Geneticists are divided about race because race is not a set of genetic facts. It is, rather, the result of a complex negotiation between genetic facts (patterns of difference) and social facts (perceptions of otherness). Looking at a Swede, a Greek, a Turk, a Moroccan, a Persian, a Pakistani, and an Ethiopian, or sequencing their DNA, doesn't tell you how many biological categories of people they represent. The categories are imposed upon the sample, not discovered within it. Race is not a fact of human biology but of nature/culture, and no amount of genetic data and statistical analysis is going to resolve it.

Worse yet, if you combine the fallacy of a reified human nature with the fallacy of reified human races, it becomes easy to see how human races could be imagined to possess their own distinct natures.

DARWIN'S VENTRILOQUISTS

It seems to me that one of the most important obstacles in the evolution-creation debate is biology's reluctance to deny the racists the claim to be speaking in Darwin's name. It isn't just the simple expediency of having the albatross of racism hung needlessly around Darwin's neck, but a broader issue that goes back to the dawn of modern science. Science was, after all, once in need of justification. Its primary purpose ostensibly is to make people's lives better. If science serves instead to make people's lives worse, then it stands to reason we should neither support it nor condone it. Anything that purposely makes people's lives worse is evil and consequently should not be acceptable to us as citizens.

How, then, do we respond to a science that tells us, in Darwin's name (as some did in the early 1960s), that schools should be racially segregated because people of one race are simply not as innately smart as people of another?[11] If the claim is false, which it seems to be, then science is poorly served by the claim and the scientists who make it should be discredited. But even if the claim were true, it would be bad for science, for it would make science little more than a tool for perpetuating social inequalities rather than for trying to ameliorate them. It would represent science as an insensate instrument for making people's lives worse—the specter of the amoral Dr. Frankenstein rather than the avuncular Dr. Einstein. Science that makes people's lives worse should not be done; *anything* that makes peoples lives worse should not be done. Good and evil may not be science, but the criteria for distinguishing between them are important for scientists to learn, and for everyone to learn—as, of course, the long-forgotten authors of the Garden of Eden story attempted to communicate.

What I am trying to set up here is that a classic idea—that there are *good* uses of evolution and *bad* uses of evolution—is simply an illusion. So, for a recent example, the biological anthropologist Rick Bribiescas writes: "Shoddy and irresponsible research tainted by political agendas has made its way into the academic mainstream. . . . But to totally discount evolutionary theory because of its previous misuse would be to outlaw matches in response to the crimes of arsonists."[12]

But arguing about its *misuses* at all presupposes that you can readily distinguish them from its *uses,* which the long and sad history of scientific racism shows that you simply cannot do. This passage presumes a classic categorical distinction between scientific ideas and their applications, or between the world of facts (on the one hand) and of values (on the other). That distinction is convenient and self-serving, because it gives scientists the wiggle room not to have to worry about the implications or the cultural context of what they say and do. By separating science from its applications, scientists are now absolved of any responsibility for the latter.

That distinction began to be untenable with the famous 1947 epigram by the father of the atomic bomb, J. Robert Oppenheimer, to the effect that "physicists have known sin." That sin was not plagiarism or avarice or heresy; it was developing and releasing an astonishing power of destruction upon the human race and the world. And neither was it an evil application; it was their job, it was their responsibility to the war effort.

The older view is a philosophical fossil from the Enlightenment, that humans are rational beings and animals are not, and that science is merely heightened rationality, free of interest conflicts or prejudice—positions that no competent anthropologist would hold today. By the mid-twentieth century it had come to be accepted that it is not so much rationality that makes us human as culture (in the anthropological and philosophical rather than in the newer ethological sense)—something locally rational, intellectual, and coherent yet to a large extent also irrational, affective, and capricious but always ultimately meaningful.

Moreover, the idea that science stands apart from culture has never really held up. Science has always required patronage, and that can come from business, from the state, or simply from the support of the masses. Science is always connected to other interests: nationalist, capitalist, populist. Science is cultural. As a human activity, science *must* be cultural; to see it as disconnected from culture is to see it as somehow nonhuman, as standing apart from the set of values, codes, and stories through which our humanity is actually constituted. It is not that culture gets in the way of science, then; it is that everything human, including science, is ipso facto cultural. As the sociologist Barry Barnes put it, scientists like to be thought of as court jester figures—oblivious to, disinterested in, and disentangled from interests of power—but in fact they bring a highly self-interested agenda to the table of public policy.[13]

To maintain that science takes place among humans interacting socially in a network of meanings and interests, and then to set science apart from that very network of meanings and interests, is to make scientific behavior nonhuman—which is paradoxical, given that the only known scientific activity is in fact carried out by humans.

ANTHROPOLOGY AND SCIENCE

If the wall between science and culture, or facts and values, is semipermeable, and that distinction from the Enlightenment is no longer sustainable, then you can no longer think of science as having products or uses that are either good or bad. Science itself is both good and bad,[14] and consequently it is the responsibility of scientists, as it is the responsibility of anybody, to appreciate the significance of that distinction and presumably to side with the good, or else to bear the consequences of failing to do so.

Right and wrong are very real and very human, even if they are learned and locally specific. The development of categories of behavior, of codes and standards, some being special or sacred or taboo, is an autapomorphic feature, a uniquely derived trait of our species, and is as effective a zoological identifier of humans among the living primates as bipedalism is. You can train a dog, and you can train a chimpanzee, but you cannot get them to remember the Sabbath Day and to keep it holy. It just makes no sense to them. Crickets chirp, kangaroos hop, and humans make, follow, and break rules. It is what we evolved doing. Obviously primates strategize and modulate their behavior, but this is different; it marks the distinction between not doing something because you don't think you can get away with it and not doing something because you just shouldn't. And it doesn't matter why—*that's just not the way we do things around here, and it just isn't right.*

So anyone who says, "I am a scientist, so I don't have to worry about right and wrong—that's somebody else's problem," is actually denying, in their own behavior, what is arguably a fundamental product of human evolution, for any activity not connected to the realm of morality and values is not human, pure and simple. For *science* to be such an activity, it would have to be undertaken by intelligent chimpanzees, or perhaps by Vulcans or androids—but not by people.

That is why scientific racism is worse than creationism. Creationism, at worst, makes (or leaves) people ignorant (of science). But scientific racism actually degrades people's lives. Like the scientific atheists, the scientific racists manage to make science look bad, which their opponents cannot do themselves. If I were a scientist, that is something I would worry about—the threat from within.

How is science possibly served by reducing social inequalities to unequal endowments and creating more impediments to the aspiration of a better life? That's not what science was supposed to be for. If Francis Bacon was wrong about science being for a better life, then either we need to rethink the terms of the original social contract with science (and presumably get rid of it) or else appreciate that science is a cultural activity and thus subject to the same limits and questions as other instruments of social power. We have to identify its conflicting interests and appreciate that all ostensibly scientific statements do not have equal truth value, and that understanding the cultural circumstances which make it so is integral to a comprehensive scientific education.

An anthropology of science reveals scientific questions to be more complex than they are often taken to be. Are genetically modified foods needed by the world or just by Monsanto? Is it scientific to support the interests of distributing the world's food to the people who need it or to support the interests of biotechnology companies in developing new kinds of foods?[15] Is it scientific to treat mental health pharmaceutically rather than socially?[16] Is it scientific to imagine that the previous question can be answered independently of the interests of the pharmaceutical industry? Is it scientific to ignore what local people think about their environment, or is that information crucial in developing successful conservation programs?[17] Is it scientific to imagine that our ideas about life are fixed and not affected by technology and society?[18] Is it scientific to imagine that something as cultural as the production of new human life could be technologized, without a complex and reciprocal interaction between the uses of the techniques and the perceptions of the users?[19] Is it scientific not to worry about the impact of science on the lives of modern citizens?[20] Is it scientific to ignore the context and agency of culture in coproducing a human being?

Francis Bacon wrote that knowledge is power. Science is both, and the relationship is clearer in Bacon's original: *scientia potestas est*.[21] The study of nature is powerful, and power is cultural. Engaging cultural issues is essential for understanding science; it is not an antithesis of science, as spirituality is. An anthropology of science is relevant to understanding the history and present status of scientific theories, to evaluating the reliability of scientific claims, to understanding human evolution and diversity, to making sense of the process by which scientific knowledge is produced and accepted, and to the effective dissemination and widespread application of scientific knowledge. An anthropology of science provides a frame for understanding science in a more comprehensive and accurate manner than is otherwise possible, and connects it to the world of human experience and social life, which ultimately is what makes science possible.

Notes

PREFACE

1. L. Nader, The three-cornered constellation: Magic, science, and religion revisited, in *Naked Science: Anthropological Inquiry into Boundaries, Power, and Knowledge*, ed. L. Nader, 259–276 (New York: Routledge, 1996).

2. D. Segal, Anthropology and/in/of science, *Cultural Anthropology* 16 (2001): 451–452.

3. A rich literature has emerged, especially among medical anthropologists. See, for example, R. Rapp, *Testing Women, Testing the Fetus: The Social Impact of Amniocentesis in America* (New York: Routledge, 2000); P. Rabinow, *French DNA: Trouble in Purgatory* (Chicago: University of Chicago Press, 1999); M. Lock, Eclipse of the gene and the return of divination, *Current Anthropology* 46 (2005): S47–S70; M. Everett, The "I" in the gene: Divided property, fragmented personhood, and the making of a genetic privacy law, *American Ethnologist* 34 (2007): 375–386; S. Palmié, Genomics, divination, "racecraft," *American Ethnologist* 34 (2007): 205–222; and G. Palsson, *Anthropology and the New Genetics* (New York: Cambridge University Press, 2007).

4. And you know who you are!

5. A. Walker and P. Shipman, *The Ape in the Tree: An Intellectual and Natural History of* Proconsul (Cambridge, MA: Harvard University Press, 2005).

6. N. Eldredge and I. Tattersall, *The Myths of Human Evolution* (New York: Columbia University Press, 1982); M. Landau, *Narratives of Human Evolution* (New Haven: Yale University Press, 1991); D. Haraway, *Primate Visions* (New York: Routledge, 1989); S. Strum and L. Fedigan, eds., *Primate Encounters: Models of Science, Gender, and Society* (Chicago: University of Chicago Press, 2000); R. Corbey, *The Metaphysics of Apes: Negotiating the Animal-Human Boundary* (New

York: Cambridge University Press, 2005); T. Turner, ed., *Biological Anthropology and Ethics: From Repatriation to Genetic Identity* (Albany: State University of New York Press, 2005).

7. B. Malinowski, Magic, religion, and science, in *Science, Religion, and Reality*, ed. J. Needham, 19–84 (London: Sheldon, 1925); S. Tambiah, *Magic, Science, Religion, and the Scope of Rationality* (New York: Cambridge University Press, 1990); S. Franklin, Science as culture, cultures of science, *Annual Review of Anthropology* 24 (1995): 163–184; P. Rabinow, *Making PCR: A Story of Biotechnology* (Chicago: University of Chicago Press, 1996); D. Haraway, *Modest_Witness@Second_Millennium: FemaleMan© Meets_OncoMouse™: Feminism and Technoscience* (New York: Routledge, 1997); G. Downey and J. Dumit, eds., *Cyborgs and Citadels: Anthropological Interventions in Emerging Sciences and Technologies* (Santa Fe, NM: School of American Research Press, 1997); J. Fujimura, Authorizing knowledge in science and anthropology, *American Anthropologist* 100 (1998): 347–360; E. Martin, Anthropology and the cultural study of science, *Science, Technology, and Human Values* 23 (1998): 24–44; R. Reid and S. Traweek, eds., *Doing Science + Culture* (New York: Routledge, 2000).

ONE. SCIENCE AS A CULTURE AND AS A "SIDE"

1. B. Malinowski, *Coral Gardens and Their Magic* (London: Allen and Unwin, 1935).

2. Academic "science wars" tend to be about the amount of science taught to students, or absorbed by them, not about the quality of the science; academic "humanities wars" tend to be about the different qualities of works taught to students, not about how much of it they get.

3. C.P. Snow, The two cultures, *New Statesman*, October 6, 1956, 413; C.P. Snow, *The Two Cultures* (New York: Cambridge University Press, 1959).

4. C.P. Snow, Two cultures, *Science* 130 (1959): 419; P.R. Gross and N. Levitt, *Higher Superstition: The Academic Left and Its Quarrels with Science* (Baltimore, MD: Johns Hopkins University Press, 1994), 7.

5. F.R. Leavis, Two cultures? The significance of C.P. Snow, *Spectator*, March 9, 1962, 297–304.

6. A. Sokal, Transgressing the boundaries: Toward a hermeneutics of quantum gravity, *Social Text* 46–47 (1996): 217–252.

7. A. Sokal, A physicist experiments with cultural studies, *Lingua Franca*, May–June 1996, 62–64.

8. J. Scott, Postmodern gravity deconstructed, slyly, *New York Times*, May 19, 1996, A1.

9. S. Weinberg, Sokal's hoax, *New York Review of Books*, August 8, 1996, 11–15.

10. D. Nelkin, What are the Science Wars really about? *Chronicle of Higher Education*, July 26, 1996, A32.

11. E. Zuckerkandl, Social constructionism, a lost cause, *Journal of Molecular Evolution* 51 (2000): 517–519.

12. E. Zuckerkandl, Perspectives in molecular anthropology, in *Classification and Human Evolution*, ed. S. L. Washburn, 243–272 (Chicago: Aldine, 1963), 247.

13. G. G. Simpson, Organisms and molecules in evolution, *Science* 146 (1964): 1535–1538.

14. With statistics, obviously. But then you risk producing results that are misleading, artifactual, or counterintuitive. For "numerical taxonomy," see D. Hull, *Science as a Process: An Evolutionary Account of the Social and Conceptual Development of Science* (Chicago: University of Chicago Press, 1988).

15. J. Marks, *What It Means to Be 98% Chimpanzee* (Berkeley: University of California Press, 2002). Five years later, the knowledge of the constructedness of this fact was so common that it could be called a "myth"; see J. Cohen, Relative differences: The myth of 1%, *Science* 316 (2007): 1836.

16. H. J. Fuller, Education or training? *Science* 120 (1954): 546.

17. C. Zirkle, Our splintered learning and the status of scientists, *Science* 121 (1955): 513–519.

18. A. Sokal and J. Bricmont, *Fashionable Nonsense: Postmodern Intellectuals' Abuse of Science* (New York: Picador, 1999).

19. J. Gleick, *Isaac Newton* (New York: Pantheon, 2003).

20. William Derham to John Conduitt, July 18, 1733, Keynes Ms. 133, King's College, Cambridge.

21. M. W. Gregory, The infectiousness of pompous prose, *Nature* 360 (1992): 11–12.

22. J. Maddox, Valediction from an old hand, *Nature* 378 (1995): 521–523.

23. Not being an anthropologist, Snow did not initially appreciate the problematic aspect of the very concept of culture in American anthropology. See A. L. Kroeber and C. Kluckhohn, Culture: A critical review of concepts and definitions, *Papers of the Peabody Museum of Archaeology and Ethnology* 47 (1952); and A. Kuper, *Culture: The Anthropologists' Account* (Cambridge, MA: Harvard University Press, 1999). Suffice it to say, we all kind of know what he meant, and what it means. See chapter 10.

24. D. J. Hess, Crosscurrents: Social movements and the anthropology of science and technology, *American Anthropologist* 109 (2007): 463–472; M. M. J. Fischer, Four genealogies for a recombinant anthropology of science and technology, *Cultural Anthropology* 22 (2007): 540–615.

TWO. THE SCIENTIFIC REVOLUTION

1. A. R. Hall, *The Scientific Revolution, 1500–1800,* 2d ed. (Boston: Beacon Press, 1962).

2. J. Barzun, *From Dawn to Decadence: 1500 to the Present* (New York: Perennial, 2000), 192–193.

3. C. Singer, Historical relations of religion and science, in *Science, Religion, and Reality,* ed. J. Needham, 85–148 (London: Sheldon, 1925).

4. F. Bacon, *The New Organon, or True Directions Concerning the Interpretation of Nature,* Aphorism 129 (1620); P. Harrison, Curiosity, forbidden knowledge, and the reformation of natural philosophy in early modern England, *Isis* 92 (2001): 265–290.

5. J. Donne, *An anatomie of the world* (1610): The first anniversary.

6. P. Gay, *The Enlightenment, an Interpretation: The Rise of Modern Paganism* (New York: Knopf, 1966); P. Gay, *The Enlightenment, an Interpretation: The Science of Freedom* (New York: Knopf, 1969).

7. M. Mersenne, *Cogitata physico-mathematica* (1644); P. Gassendi, *Syntagma philosophicum* (1658); R. Boyle, *The Sceptical Chymist* (1661); M. J. Osler, *Divine Will and the Mechanical Philosophy* (New York: Cambridge University Press, 2005).

8. M. F. A. Montagu, Edward Tyson, M.D., F.R.S., 1650–1708 and the rise of human and comparative anatomy in England, *Memoirs of the American Philosophical Society* 20 (1943).

9. B. Spinoza, *Tractatus Theologico-Politicus* (1670), chap. 12; M. Stewart, *The Courtier and the Heretic: Leibniz, Spinoza, and the Fate of God in the Modern World* (New York: W. W. Norton, 2006).

10. N. Wade, Genetic code of human life is cracked by scientists, *New York Times,* June 27, 2000.

11. Anonymous, We are guarded by spirits, declares Dr. A. R. Wallace: The great scientist who discovered the evolution theory simultaneously with Darwin tells how "superior beings control mankind," *New York Times,* October 8, 1911; Anonymous, Biologist denies chance evolution: Dr. Lucien Cuentot *[sic]* says insects' mechanical devices evidently result from design. Cuttlefish an example. Eugenics Congress listens to the theories of an eminent French scientist, *New York Times,* September 24, 1921.

12. A. Novikoff, The concept of integrative levels and biology, *Science* 101 (1945): 209–215.

13. H. Anderson, The history of reductionism versus holistic approaches to scientific research, *Endeavour* 25, no. 4 (2001): 153–156.

14. Indeed, "called"—his real name was Philippus Aureolus Theophrastus Bombastus von Hohenheim.

15. S. Shapin and S. Schaffer, *Leviathan and the Air Pump: Hobbes, Boyle and the Experimental Life* (Princeton, NJ: Princeton University Press, 1985).

16. A.R. Hall, *From Galileo to Newton* (New York: Harper, 1963).

17. R. Boyle, A defense of the doctrine touching the spring and weight of the air. In *New Experiments Physico-Mechanicall, Touching the Spring of the Air, and Its Effects (Made for the most part, in a New Pneumatical Engine)* (1662) .

18. W.J. Ashworth, Metrology and the state: Science, revenue, and commerce, *Science* 306 (2004): 1314–1317.

19. J. Gleick, *Isaac Newton* (New York: Pantheon, 2003).

20. J.D. Bernal, *Science in History,* vol. 2: *The Scientific and Industrial Revolutions* (Cambridge, MA: MIT Press, [1954] 1971).

21. T. Nagel, *The View from Nowhere* (New York: Oxford University Press, 1986).

22. S. Shapin, *The Scientific Revolution* (Chicago: University of Chicago Press, 1996).

23. J. Ray, *The Wisdom of God Manifested in the Works of Nature* (1691); W. Paley, *Natural Theology* (1802).

24. R. Nisbet, *History of the Idea of Progress* (New York: Basic Books, 1980).

25. R. Dawkins, *The Blind Watchmaker: Why the Evidence of Evolution Reveals a Universe without Design* (New York: W.W. Norton, 1986); R. Dawkins, *Unweaving the Rainbow: Science, Delusion and the Appetite for Wonder* (New York: Mariner Books, 2000).

26. J.H. Brooke, *Science and Religion: Some Historical Perspectives* (New York: Cambridge University Press, 1991).

THREE. NORMATIVE SCIENCE

1. F.J. Sulloway, Darwin's conversion: The Beagle voyage and its aftermath, *Journal of the History of Biology* 15 (1982): 325–396.

2. K. Popper, *The Logic of Scientific Discovery* [Logik der Forschung] (New York: Basic Books, [1934] 1959). For a different approach to a similar conclusion, see J. Ioannidis, Why most published research findings are false, *PLoS Medicine* 2 (2005): 0696–0701.

3. T. Kuhn, *The Structure of Scientific Revolutions* (Chicago: University of Chicago Press, 1962); D.L. Hull, A revolutionary philosopher of science, *Nature* 382 (1996): 203–204.

4. A.O. Lovejoy, *The Great Chain of Being* (Cambridge, MA: Harvard University Press, 1936).

5. These four categories were formalized by Cuvier a generation after Linnaeus.

6. Georges-Louis Leclerc, Comte de Buffon, L'Asne, in *Histoire naturelle, générale et particulière,* vol. 4, 382 (Paris: L'Imprimerie royale, 1753).

7. Except for the brief war of the biometricians and the Mendelians in England.

8. J. Sapp, The struggle for authority in the field of heredity, 1900–1932: New perspectives on the rise of genetics, *Journal of the History of Biology* 16, no. 3 (1983): 311–342; P. Bowler, *The Mendelian Revolution* (Baltimore, MD: Johns Hopkins University Press, 1989); R. Falk, The struggle of genetics for independence, *Journal of the History of Biology* 28 (1995): 219–246.

9. *Linkage,* the physical contiguity of two genes on the same chromosome. *Crossing over,* the reciprocal exchange of linked genetic variations into new combinations. *Mitochondrial inheritance,* the transmission of a small group of genes exclusively through the egg rather than equally through egg and sperm. *Pleiotropy,* the diverse physiological effects of the action of a single gene. *Epistasis,* the nonadditive effects of several genes working together. *Imprinting,* the unequal expression of genes transmitted through sperm or egg. *Codominance,* the simultaneous expression of both genetic variants in a single individual rather than one variant suppressing the other. *Quantitative traits,* those features that vary not so much in an all-or-nothing fashion (as tall or short pea plants, and wrinkled or round peas), but on a continuous scale.

10. T.H. Morgan, A.H. Sturtevant, H.J. Muller, and C.B. Bridges, *The Mechanism of Mendelian Inheritance* (New York: Henry Holt, 1915); T.H. Morgan, *A Critique of the Theory of Evolution* (Princeton, NJ: Princeton University Press, 1916).

11. C. Darwin, Pangenesis, *Nature* 3 (1871): 502–503; F. Galton, Pangenesis, *Nature* 4 (1871): 5–6.

12. T. Dobzhansky, A review of some fundamental concepts and problems of population genetics, *Cold Spring Harbor Symposium on Quantitative Biology* 20 (1955): 1–15; R.C. Lewontin, *The Genetic Basis of Evolutionary Change* (New York: Columbia University Press, 1974).

13. R. Dawkins, *River out of Eden: A Darwinian View of Life* (New York: Basic Books, 1995).

14. S. Jasanoff, Knowledge elites and class war, *Nature* 401 (1999): 531.

15. D. Danielson, Scientist's birthright, *Nature* 410 (2001): 1031.

16. T. Theocharis and M. Psimopoulos, Where science has gone wrong, *Nature* 329 (1987): 595–598; G. Holton, *Science and Anti-Science* (Cambridge, MA: Harvard University Press, 1993); K. Gottfried and K.G. Wilson, Science as a cultural construct, *Nature* 386 (1997): 545–547.

17. S. Kühl, *The Nazi Connection* (New York: Oxford University Press, 1994); F.A. Woods, *The Passing of the Great Race,* 2d ed. (review), *Science* 48 (1918): 419–420; F.A. Woods, A review of reviews of Madison Grant's *Passing of the Great Race, Journal of Heredity* 14 (1923): 93–95.

18. Aleš Hrdlička to Irving Fisher, November 27, 1923, Hrdlička Papers, National Anthropological Archives, Smithsonian Institution; Earnest Hooton to

Madison Grant, November 3, 1933, Earnest A. Hooton Papers, Peabody Museum, Harvard University.

19. G.E. Allen, The Eugenics Record Office at Cold Spring Harbor, 1910–1940, *Osiris* 2 (1986): 225–264; B. Glass, Geneticists embattled: Their stand against rampant eugenics and racism in America during the 1920s and 1930s, *Proceedings of the American Philosophical Society* 130 (1986): 130–154.

20. H.J. Muller, The dominance of economics over eugenics, *Scientific Monthly* 37 (1933): 40–47; L. Hogben, *Genetic Principles in Medicine and Social Science* (London: Williams and Norgate, 1931).

21. D.J. Kevles, *In the Name of Eugenics* (Berkeley: University of California Press, 1985).

22. J.V. Neel and W.J. Schull, *Human Heredity* (Chicago: University of Chicago Press, 1954), 337.

23. E. Sinnott and L. C. Dunn, *Principles of Genetics* (New York: McGraw-Hill, 1925), 406.

24. R.N. Proctor, *Value-Free Science? Purity and Power in Modern Knowledge* (Cambridge, MA: Harvard University Press, 1991).

25. H. Collins and T. Pinch, *The Golem: What Everyone Should Know about Science* (New York: Cambridge University Press, 1993).

26. V. Bush, Summary of the report to the president on a program for postwar scientific research by Vannevar Bush, director of OSRD, *Science* 102 (1945): 79–81.

27. N.W. Pirie, The maldistribution of research effort, in *The Science of Science*, ed. M. Goldsmith and A. Mackay, 198–213 (London: Pelican, 1964).

28. R.K. Merton, Science and technology in a democratic order, *Journal of Legal and Political Sociology* 1 (1942): 115–126, later retitled The normative structure of science, in R.K. Merton, *The Sociology of Science: Theoretical and Empirical Investigations* (Chicago: University of Chicago Press, 1973); D.A. Hollinger, *Science, Jews, and Secular Culture* (Princeton, NJ: Princeton University Press, 1996).

29. On whether the NSF should serve the needs of the nation or science first, see D.J. Kevles, The National Science Foundation and the debate over postwar research policy, 1942–1945, *Isis* 68 (1977): 5–26; C. Mukerji, *A Fragile Power: Scientists and the State* (Princeton, NJ: Princeton University Press, 1989).

On how closely the NSF should be overseen, see M. Polanyi, The autonomy of science, *Scientific Monthly* 60 (1945): 141–150; L. Kartman, Soviet genetics and the "autonomy of science," *Scientific Monthly* 61 (1945): 67–70.

On the proper relationship between a scientist and the financial interests of patents, see W. Davis, Scientists divided, *Science* 103 (1946): 688.

30. D. King, The politics of social research: Institutionalizing public funding regimes in the United States and Britain, *British Journal of Political Science* 28 (1998): 415–444.

FOUR. SCIENCE AS PRACTICE

1. T. Ingold, *Lines: A Brief History* (London: Routledge, 2007).

2. The idea that linear stratification by wealth is also natural and acceptable was challenged in various ways in the nineteenth and twentieth centuries but remains far more firmly entrenched and unquestioned in the United States.

3. Of course, as a bit of cultural knowledge, the equation is well-nigh ubiquitous.

4. Christy Mathewson of the New York Giants was the preeminent pitcher in 1905, when Einstein published his paper on special relativity.

5. R. Benedict, *Patterns of Culture* (New York: Houghton Mifflin, 1934); B.H. Smith, *Scandalous Knowledge: Science, Truth and the Human* (Edinburgh: Edinburgh University Press, 2005).

6. Half a century later, the Australian anthropologist Derek Freeman, attempting to defend the thesis that patterns of adolescent sexuality are innate (in the name of sociobiology), argued that Mead's fieldwork was flawed and biased and her conclusions inaccurate. While Freeman's work is still often brandished uncritically by evolutionary psychologists, the judgment of anthropologists, and especially anthropologists of Samoa, is that Freeman's interpretation was wronger than Mead's. See chapter 6.

7. M.A. Largent, *Breeding Contempt: The History of Coerced Sterilization in the United States* (New Brunswick, NJ: Rutgers University Press, 2008).

8. This historical relativist critique of science is associated with the Edinburgh school, or strong program, of science studies. B. Barnes, *Scientific Knowledge and Sociological Theory* (London: Routledge, 1974).

9. R. Lewin, *Bones of Contention: Controversies in the Search for Human Origins* (New York: Simon and Schuster, 1987); J.S. Weiner, *The Piltdown Forgery* (London: Oxford University Press, 1955); M. McCarty, *The Transforming Principle: Discovering That Genes Are Made of DNA* (New York: W.W. Norton, 1985).

10. Ian Parker, Richard Dawkins's evolution: An irascible don becomes a surprising celebrity, *New Yorker*, September 9, 1996; R. Dawkins, *River out of Eden* (New York: Basic Books, 1995).

11. A bit uglier precedent involved segregationist activists in the 1960s making the argument for the intellectual inferiority of blacks on cultural grounds, in spite of whatever left-wing nonsense anthropologists had to say about it. But as the anthropologist Stanley Diamond pointed out (*Science* 135 [1962]: 961), what other scientific authorities would you want to consult in order to make cultural comparisons? It is, after all, the anthropologist who "deals with the origin and growth of cultural behavior and with the cultural interaction of human groups." The convergence of Richard Dawkins's views with those of the segregationists is ample testimony to the crude anti-intellectualism at the heart of the argument.

12. The political journalist Thomas Friedman, however, tried to reduce cultural diversity to a simple pro-globalization and anti-globalization metaphor in *The Lexus and the Olive Tree* (New York: Farrar, Straus, Giroux, 1999).

13. J. Lawrence and R. E. Lee, *Inherit the Wind* (New York: Random House, 1955), Act II, Scene 2.

14. H. Collins and T. Pinch, *The Golem: What Everyone Should Know about Science* (New York: Cambridge University Press, 1993).

15. Also: shills, cheerleaders, propagandists, fanatics, extremists, and right-wing lug nuts.

16. G. Holton, Subelectrons, presuppositions, and the Millikan-Ehrenhaft dispute, *Historical Studies in the Physical Sciences* 9 (1978): 161–224; A. Franklin, *The Neglect of Experiment* (New York: Cambridge University Press, 1986); D. Goodstein, In defense of Robert Andrews Millikan, *Engineering and Science* 4 (2000): 30–38.

17. P. Medawar, Is the scientific paper a fraud? in *The Threat and the Glory*, 228–234 (New York: Oxford University Press, [1963] 1991).

18. S. Shapin, The art of persuasion, *Nature* 448 (2007): 751–752.

19. B. Barnes, *Interests and the Growth of Knowledge* (London: Routledge and Kegan Paul, 1977).

20. H. Gusterson, *Nuclear Rites: A Weapons Laboratory at the End of the Cold War* (Berkeley: University of California Press, 1996) .

21. E.R. Leach, Don't say "boo" to a goose, *New York Review of Books* 7 (December 15, 1966); D. Haraway, *Primate Visions* (New York: Routledge, 1989); S. Sperling, Baboons with briefcases: Feminism, functionalism, and sociobiology in the evolution of primate gender, *Signs* 17 (1991): 1–27; S.C. Strum and L.M. Fedigan, *Primate Encounters: Models of Science, Gender, and Society* (Chicago: University of Chicago Press, 2000); A. Rees, Anthropomorphism, anthropocentrism, and anecdote: Primatologists on primatology, *Science, Technology and Human Values* 26 (2001): 227–247; D. Hart and R. Sussman, *Man the Hunted: Primates, Predators, and Human Evolution* (New York: Westview Press, 2005).

22. P. Asquith, Japanese science and Western hegemonies: Primatology and the limits set to questions, in *Naked Science: Anthropological Inquiry into Boundaries, Power, and Knowledge*, ed. L. Nader, 239–258 (New York: Routledge, 1996).

23. G.P. Murdock, British social anthropology, *American Anthropologist* 53 (1951): 465–473; A.R. Radcliffe-Brown, Historical note on British social anthropology, *American Anthropologist* 54 (1952): 275–277; B. Halstead, Anti-Darwinian theory in Japan, *Nature* 317 (1985): 587–589; N. Hokkyo, Anti-Darwinism in Japan, *Nature* 322 (1986): 107.

24. R.N. Proctor, "Everyone knew but no one had proof": Tobacco industry use of medical history expertise in US courts, 1990–2002, *Tobacco Control* 15 (2006): iv117–iv125.

25. R. Horton, The dawn of McScience, *New York Review of Books* 51 (March 11, 2004): 7–9.

26. G. Kolata, Hope in the lab, *New York Times,* May 3, 1998; F. Barringer, Cancer drug news puts a focus on reporters and book deals, *New York Times,* May 8, 1998; E. Marshall, The power of the front page of the *New York Times, Science* 280 (1998): 996–997; A. Gawande, Mouse hunt, *New Yorker,* May 18, 1998, 5–6; G. Huberman and T. Regev, Contagious speculation and a cure for cancer: A nonevent that made stock prices soar, *Journal of Finance* 56 (2001): 387–396.

27. P.W. Huber, *Galileo's Revenge: Junk Science in the Courtroom* (New York: Basic Books, 1991).

28. J.D. Watson, *The Double Helix: A Personal Account of the Discovery of the Structure of DNA* (New York: Atheneum, 1968), ix, 51, 106, 114.

29. A. Sayre, *Rosalind Franklin and DNA* (New York: Norton, 1975); B. Maddox, *Rosalind Franklin: The Dark Lady of DNA* (New York: HarperCollins, 2002).

30. A.F. Richard, *Primates in Nature* (San Francisco: W.H. Freeman, 1985); B. Smuts, D. Cheney, R. Seyfarth, and R. Wrangham, eds., *Primate Societies* (Chicago: University of Chicago Press, 1987); C. Campbell, A. Fuentes, K. MacKinnon, M. Panger, and S. Bearder, eds., *Primates in Perspective* (New York: Oxford University Press, 2006); K.B. Strier, *Primate Behavioral Ecology* (New York: Allyn and Bacon, 2006).

31. N. Isaac, J. Mallet, and G. Mace, Taxonomic inflation: Its influence on macroecology and conservation, *Trends in Ecology and Evolution* 19 (2004): 464–469; I. Tattersall, Madagascar's lemurs: Cryptic diversity or taxonomic inflation? *Evolutionary Anthropology* 16 (2007): 12–23; J. Marks, Anthropological taxonomy as both subject and object: The consequences of descent from Darwin and Durkheim, *Anthropology Today* 23, no. 4 (2007): 7–12.

FIVE. THE PROBLEM OF CREATIONISM

1. L. Godfrey, ed., *Scientists Confront Creationism* (New York: W.W. Norton, 1984); A. Montagu, ed., *Science and Creationism* (New York: Oxford University Press, 1984); E. Scott, *Evolution vs. Creationism: An Introduction* (Berkeley: University of California Press, 2005); M. Ruse, *The Evolution-Creation Struggle* (Cambridge, MA: Harvard University Press, 2005).

2. The first blast of anti-Darwinian reaction came from the last natural theologians, Richard Owen in England and Louis Agassiz in the United States, who could not conceive of natural history independent of the hand of God but had no coherent alternative theory. Likewise, Alfred Russel Wallace, widely known as the codiscoverer of natural selection, believed that since savages have brains equal to civilized people, but don't use them fully, that organ must have arisen

by some process other than natural selection. By the end of his life he was going on about life being governed by spirits wiser than ourselves.

3. Thus did kings become kings: Magic, art, and their evolution as set forth in Dr. Frazer's "Golden Bough," *New York Times,* August 20, 1911.

4. H. L. Mencken, *A Mencken Chrestomathy* (New York: Knopf, 1949), 624.

5. H. F. Osborn, Hesperopithecus, the first anthropoid primate found in America, *Proceedings of the National Academy of Sciences, USA* 8 (1922): 245–246.

6. W. K. Gregory, Hesperopithecus apparently not an ape nor a man, *Science* 66 (1927): 579–581.

7. Yearning mountaineers' souls need reconversion nightly, Mencken finds, *Baltimore Evening Sun,* July 13, 1925; Battle now over, Mencken sees; Genesis triumphant and ready for new jousts, *Baltimore Evening Sun,* July 18, 1925.

8. E. J. Larson, *Summer for the Gods: The Scopes Trial and America's Continuing Debate over Science and Religion* (New York: Basic Books, 1997); R. H. Robbins, William Jennings Bryan and the trial of John T. Scopes, in *Darwin and the Bible: The Cultural Confrontation* (New York: Allyn and Bacon, 2009). It would not be until 1968 (in *Epperson v. Arkansas*) that the Supreme Court declared banning the teaching of evolution to be unconstitutional.

9. J. Huxley, "At Random": A television preview, in *Evolution after Darwin,* vol. 3: *Issues in Evolution,* ed. S. Tax, 41–65 (Chicago: University of Chicago Press, 1960).

10. J. V. Grabiner and P. D. Miller, Effects of the Scopes trial, *Science* 185 (1974): 832–837.

11. W. R. Overton, Creationism in schools: The decision in McLean versus the Arkansas Board of Education, *Science* 215 (1982): 934–943.

12. Ibid.

13. *Edwards v. Aguillard,* 482 U.S. 578 (1987).

14. B. Forrest and P. R. Gross, *Creationism's Trojan Horse: The Wedge of Intelligent Design* (New York: Oxford University Press, 2004).

15. P. E. Johnson, *Objections Sustained: Subversive Essays on Evolution, Law, and Culture* (Downer's Grove, IL: InterVarsity Press, 1998); M. J. Behe, *Darwin's Black Box: The Biochemical Challenge to Evolution* (New York: Free Press, 1996); M. J. Behe, Darwin under the microscope, *New York Times,* October 29, 1996.

16. W. A. Dembski, *No Free Lunch: Why Specified Complexity Cannot Be Purchased without Intelligence* (Lanham, MD: Rowman and Littlefield, 2002); J. Wells, *Icons of Evolution* (New York: Regnery, 2000).

17. E. Bumiller, Bush remarks roil debate over teaching of evolution, *New York Times,* August 3, 2005; C. Schonborn, Finding design in nature, *New York Times,* July 5, 2005.

18. S. Gaukroger, *The Emergence of a Scientific Culture: Science and the Shaping of Modernity, 1210–1685* (New York: Oxford University Press, 2007); R. Olson, *Sci-*

ence Deified and Science Defied: The Historical Significance of Science in Western Culture, vol. 2: *From the Early Modern Age through the Early Romantic Era, ca. 1640 to ca. 1820* (Berkeley: University of California Press, 1990).

19. N. Wade, Genetic code of human life is cracked by scientists, *New York Times*, June 27, 2000.

20. K. Thomson, *Before Darwin: Reconciling God and Nature* (New Haven, CT: Yale University Press, 2005).

21. R. L. Numbers, *The Creationists* (New York: Knopf, 1992).

22. F. Odling-Smee, K. Laland, M. Feldman, and L. Schaaf, *Niche Construction: The Neglected Process in Evolution* (Princeton, NJ: Princeton University Press, 2003); J. Turner, *The Tinker's Accomplice: How Design Emerges from Life Itself* (Cambridge, MA: Harvard University Press, 2007); William Shakespeare, *Hamlet* (1603), Act III, Scene 2.

23. And I'm not even particularly religious.

24. R. Pearl, The First International Eugenics Congress, *Science* 36 (1912): 395–396.

25. H. F. Osborn, The Second International Congress of Eugenics: Address of welcome, *Science* 54 (1921): 311–313.

26. C. B. Davenport, Research in eugenics, *Science* 54 (1921): 391–397.

27. G. K. Chesterson, *Eugenics and Other Evils* (London: Cassell, 1922), 39.

28. C. Darrow, The Edwardses and the Jukeses, *American Mercury* 6 (1925): 147–157; C. Darrow, The eugenics cult, *American Mercury* 8 (1926): 129–137. On Goddard's response to inquiries about the name of the feebleminded tavern girl, see H. H. Goddard, In defense of the Kallikak study, *Science* 95 (1942): 574–576; J. Smith, *Minds Made Feeble: The Myth and Legacy of the Kallikaks* (Rockville, MD: Aspen Systems, 1985).

29. H. L. Mencken, On eugenics, *Baltimore Sun*, May 15, 1927.

30. *San Francisco Bulletin*, October 24, 1927; *Salt Lake City Telegram*, October 25, 1927 (from the papers of Leon F. Whitney).

31. Matthew 24:5; Frederick Engels to Edouard Bernstein, November 3, 1882.

32. S. Satel, I am a racially profiling doctor, *New York Times*, May 5, 2002; A. F. Leroi, A family tree in every gene, *New York Times*, March 14, 2005; T. Duster, Race and reification in science, *Science* 307 (2005): 1050–1051; J. Kahn, Patenting race, *Nature Biotechnology* 24 (2006): 1349–1351; J. Marks, Long shadow of Linnaeus's human taxonomy, *Nature* 447 (2007): 28.

33. R. Dawkins, *The God Delusion* (New York: Bantam, 2006).

34. P. Cohen, A split emerges as conservatives discuss Darwin, *New York Times*, May 5, 2007.

35. J. Miller, E. Scott, and S. Okamoto, Public acceptance of evolution, *Science* 313 (2006): 765–766.

36. Quantum electrodynamics is an esoteric field of physics with the principal distinction of having been invented by Richard Feynman. Helminthology is the study of parasitic worms, generally identified in feces. Need I say more? And scientists writing about theology generally produce material as inane as theologians writing about science. You might as well listen to what your plumber says about both subjects.

37. R. Lowie, Religion in human life, *American Anthropologist* 65 (1963): 532–542.

38. C.P. Toumey, *God's Own Scientists: Creationists in a Secular World* (New Brunswick, NJ: Rutgers University Press, 1994).

39. H.L. Mencken, Caveat against science, *American Mercury* 12 (1927): 126 127; reprinted in Mencken, *A Mencken Chrestomathy*, 330–333.

40. G. Holton, *Science and Anti-Science* (Cambridge, MA: Harvard University Press, 1993); P. Gross and N. Levitt, *Higher Superstition: The Academic Left and Its Quarrels with Science* (Baltimore, MD: Johns Hopkins University Press, 1994).

41. If, by "monkeys," we refer to arboreal, tailed, quadrupedal anthropoids.

SIX. BOGUS SCIENCE

1. J.R. Powell, Reviewing misconduct? *American Scientist* 81 (1993): 408.

2. M.J. Kottler, From 48 to 46: Cytological technique, preconception, and the counting of human chromosomes, *Bulletin of the History of Medicine* 48 (1974): 465–502; D.A. Hungerford, Some early studies of human chromosomes, 1879–1955, *Cytogenetics and Cell Genetics* 20 (1978): 1–11.

3. H. de Winiwarter, Études sur la spermatogenèse humaine, *Archives de Biologie* 27 (1912): 93, 147–149; H. de Winiwarter, Le nombre des chromosomes chez l'homme, *Revue Anthropologique* 30 (1920): xxv–xxxii; M.F. Guyer, A note on the accessory chromosomes of man, *Science* 39 (1914): 721–722; T.H. Morgan, Has the white man more chromosomes than the Negro? *Science* 39 (1914): 827–828.

4. T.S. Painter, The Y-chromosome in mammals, *Science* 53 (1921): 503–504; T.S. Painter, Studies in mammalian spermatogenesis. II. The spermatogenesis of man, *Journal of Experimental Zoology* 37 (1923): 291–334; T.S. Painter, The sex chromosomes of man, *American Naturalist* 58 (1924): 506–524; T.S. Painter, Chromosome numbers in mammals, *Science* 61 (1925): 423–424. This paper also records forty-eight chromosomes in *Macaca*, but the actual number is forty-two. E.G. Conklin, *Heredity and Environment in the Development of Men*, 4th ed. (Princeton, NJ: Princeton University Press, 1922), 166. The quotation is from Kottler, From 48 to 46.

5. T.C. Hsu, Mammalian chromosomes in vitro. I. The karotype of man, *Journal of Heredity* 43 (1952): 167–172; C.D. Darlington and A. Haque, Chromosomes of monkeys and men, *Nature* 175 (1955): 32.

6. J.H. Tjio and A. Levan, *Hereditas* 42 (1956): 1–6; C.E. Ford and J.L. Hamerton, The chromosomes of man, *Nature* 178 (1956): 1020–1023.

7. T.C. Hsu, *Human and Mammalian Cytogenetics: An Historical Perspective* (New York: Springer-Verlag, 1979).

8. J. Giles, The trouble with replication, *Nature* 442 (2006): 344–347; E. Check, Stem cells: The hard copy, *Nature* 446 (2007): 485–486.

9. S. Zuckerman, South African fossil anthropoids, *Nature* 165 (1950): 652; W. Le Gros Clark, South African fossil hominoids, *Nature* 165 (1950): 893–894; W.E. Le Gros Clark, South African fossil hominoids, *Nature* 166 (1950): 791–792; S. Zuckerman, South African fossil hominoids, *Nature* 166 (1950): 158–159; J. Bronowski and W. Long, Statistical methods in anthropology, *Nature* 168 (1951): 794; E. Ashton and S. Zuckerman, Statistical methods in anthropology, *Nature* 168 (1951): 1117–1118.

10. See below for the homeopathy controversy. J. Benveniste, B. Ducot, and A. Spira, Memory of water revisited, *Nature* 370 (1994): 322.

11. D. Schneider, The coming of a sage to Samoa, *Natural History* 92, no. 6 (1983): 4–10; A. Kuper, Coming of age in anthropology? *Nature* 338 (1989): 453–455; M. Orans, Mead misrepresented, *Science* 283 (1999): 1649–1650; P. Monaghan, An Australian historian puts Margaret Mead's biggest detractor on the psychoanalytic sofa, *Chronicle of Higher Education* 52, no. 19 (2006): A14.

12. S.L. Washburn, The Piltdown hoax: Piltdown 2, *Science* 203 (1979): 955–958.

13. H.F. Judson, *The Great Betrayal* (Orlando, FL: Harcourt, 2004), 82.

14. A. McLaren, A prehistory of the social sciences: Phrenology in France, *Comparative Studies in Society and History* 23 (1981): 3–22; R. Cooter, *The Cultural Meaning of Popular Science: Phrenology and the Organization of Consent in Nineteenth-Century Britain* (New York: Cambridge University Press, 1984); E.A. Spitzka, Review of *Phrenology or the Doctrine of the Mental Phenomena* by J.G. Spurheim, *Science* 30 (1909): 310–311; S. Franz, New phrenology, *Science* 35 (1912): 321–328.

15. M. Iacoboni, J. Freedman, J. Kaplan, K. Jamieson, T. Freedman, B. Knapp, and K. Fitzgerald, This is your brain on politics, *New York Times,* November 11, 2007; A. Aron et al., Politics and the brain, *New York Times,* November 14, 2007; G. Miller, Growing pains for fMRI, *Science* 320 (2008): 1412–1414.

16. F.A. Woods, *The Passing of the Great Race,* 2d ed. (review), *Science* 48 (1918): 419–420; F.A. Woods, A review of reviews of Madison Grant's *Passing of the Great Race, Journal of Heredity* 14 (1923): 93–95; J. Huxley and A.C. Haddon, *We Europeans* (New York: Harper and Brothers, 1936); J. Spiro, *Defending the Master Race: Conservation, Eugenics, and the Legacy of Madison Grant* (Burlington: University of Vermont Press, 2009).

17. *The Passing of the Great Race* was reviewed favorably in *Science* and a bit

skeptically in *Nature* by the anatomist Arthur Keith. But simply to get noticed by the two most prestigious science journals in the world is significant.

18. Anonymous, Reich opens race study: Halle University course said to be based on American models, *New York Times,* August 2, 1933, 6.

19. Madison Grant, 71, zoologist, is dead, *New York Times,* May 31, 1937, 19. Anthropologists by and large repudiated Grant's work, but who ever listens to them?

20. E. Davenas, F. Beauvais, J. Amara, M. Oberbaum, B. Robinzon, A. Miadonnai, A. Tedeschi, B. Pomeranz, P. Fortner, and P. Belon, Human basophil degranulation triggered by very dilute antiserum against IgE, *Nature* 333 (1988): 816–818; J. Maddox, J. Randi, and W. Stewart, "High-dilution" experiments a delusion, *Nature* 334 (1988): 287–291; J. Maddox, Maddox on the "Benveniste affair," *Science* 241 (1988): 1585–1586; C. Picart, Scientific controversy as farce: The Benveniste-Maddox counter trials, *Social Studies of Science* 24 (1994): 7–37.

21. E. Ernst, T. Saradeth, and K. Resch, Drawbacks of peer review, *Nature* 363 (1993): 296; J. Armstrong, We need to rethink the editorial role of peer reviewers, *Chronicle of Higher Education* 43 (1996): B3–B4; E. Marshall, Suit alleges misuse of peer review, *Science* 270 (1995): 1912–1914; T. Gura, Scientific publishing: Peer review, unmasked, *Nature* 416 (2002): 258–260; D. Adam and J. Knight, Publish, and be damned, *Nature* 419 (2002): 772–776.

22. R. Park, The seven warning signs of bogus science, *Chronicle of Higher Education* 49 (January 31, 2003): B20. Park's seven warning signs: (1) The discoverer pitches the claim directly to the media. (2) The discoverer says that a powerful establishment is trying to suppress his or her work. (3) The scientific effect involved is always at the very limit of detection. (4) Evidence for a discovery is anecdotal. (5) The discoverer says a belief is credible because it has endured for centuries. (6) The discoverer has worked in isolation. (7) The discoverer must propose new laws of nature to explain an observation.

23. T. H. Huxley, *Vestiges of the Natural History of Creation* (review), *British and Foreign Medico-chirurgical Review* 13 (1854): 425.

24. E.R. Leach, Don't say "boo" to a goose, *New York Review of Books* 7 (December 15, 1966).

25. J. Benveniste, Dr. Jacques Benveniste replies, *Nature* 334 (1988): 291.

26. Z. Medvedev, *The Rise and Fall of T. D. Lysenko,* trans. I. M. Lerner (New York: Columbia University Press, 1969); N. Roll-Hansen, *The Lysenko Effect: The Politics of Science* (New York: Humanity Books, 2005).

27. P. Jacobs, M. Brunton, M. Melville, R. Brittain, and W. McClemont, Aggressive behavior, mental subnormality and the XYY male, *Nature* 208 (1965): 1351–1352; E. Hook, Behavioral implications of the human XYY genotype, *Science* 179 (1973): 139–150; H. A. Witkin, S. A. Mednick, F. Schulsinger, E. Bakkestrom, K.O. Christiansen, D.R. Goodenough, K. Hirschhorn, C. Lundsteen, D.R. Owen,

J. Philip, D.R. Rubin, and M. Stocking, Criminality in XYY and XXY men, *Science* 193 (1976): 547–555.

28. P. Evans, J. Anderson, E. Vallender, S. Gilbert, C. Malcom, S. Dorus, and B. Lahn, Adaptive evolution of ASPM, a major determinant of cerebral cortical size in humans, *Human Molecular Genetics* 13 (2004): 489–494; N. Mekel-Bobrov, S. Gilbert, P. Evans, E. Vallender, J. Anderson, R. Hudson, S. Tishkoff, and B. Lahn, Ongoing adaptive evolution of ASPM, a brain size determinant in *Homo sapiens, Science* 309 (2005): 1720–1722; P. Evans, S. Gilbert, N. Mekel-Bobrov, E. Vallender, J. Anderson, L. Vaez-Azizi, S. Tishkoff, R. Hudson, and B. Lahn, Microcephalin, a gene regulating brain size, continues to evolve adaptively in humans, *Science* 309 (2005): 1717–1720; M. Balter, Brain man makes waves with claims of recent human evolution, *Science* 314 (2006): 1871; M. Currat, L. Excoffier, W. Maddison, S. Otto, N. Ray, M. Whitlock, and S. Yeaman, Comment on "Ongoing adaptive evolution of ASPM, a brain size determinant in *Homo sapiens*" and "Microcephalin, a gene regulating brain size, continues to evolve adaptively in humans," *Science* 313 (2006): 172a–172b; R. Woods, N. Freimer, J. De Young, S. Fears, and N. Sicotte, Normal variants of microcephalin and ASPM do not account for brain size variability, *Human Molecular Genetics* 15 (2006): 2025–2029; A. Regalado, Scientist's study of brain genes sparks a backlash, *Wall Street Journal*, June 16, 2006, A12; C.E. Atkins, Seed interview: Bruce Lahn, *Seed Magazine*, http://seedmagazine.com, September 11, 2006.

29. G. Taubes, *Bad Science: The Short Life and Weird Times of Cold Fusion* (New York: Random House, 1993).

30. G. Kolata, *Clone: The Road to Dolly and the Path Ahead* (New York: William Morrow, 1997).

31. D. Hamer, S. Hu, V. Magnuson, N. Hu, and A. Pattatucci, A linkage between DNA markers on the X chromosome and male sexual orientation, *Science* 261 (1993): 321–327; A. Fausto-Sterling and E. Balaban, Genetics and male sexual orientation, *Science* 261 (1993): 1257; J. Maddox, Sexual orientations, *Nature* 365 (1993): 702; D. Hamer and P. Copeland, *The Science of Desire: The Search for the Gay Gene and the Biology of Behavior* (New York: Simon and Schuster, 1994); S. Hu, A. Pattatucci, C. Patterson, L. Li, D. Fulker, S. Cherny, L. Kruglyak, and D. Hamer, Linkage between sexual orientation and chromosome Xq28 in males but not in females, *Nature Genetics* 11 (1995): 248–256; E. Marshall, NIH's "gay gene" study questioned, *Science* 268 (1995): 1841; D. Hamer and P. Copeland, *Living with Our Genes* (New York: Anchor, 1998); G. Rice, N. Risch, and G. Ebers, Genetics and male sexual orientation, *Science* 285 (1999): 803a; D. Hamer, *The God Gene* (New York: Doubleday, 2004).

32. S. Silverman and S. McKinnon, *Complexities: Beyond Nature and Nurture* (Chicago: University of Chicago Press, 2005).

33. I. Velikovsky, *Worlds in Collision* (New York: Doubleday, 1950); H. Bauer,

Beyond Velikovsky: The History of a Public Controversy (Urbana: University of Illinois Press, 1999).

34. Velikovsky seems to have been inspired here by a speculative 1883 book called *Ragnarok*, which claimed that geologists were all wrong about the history of the earth; the planet had been struck by a comet in the distant past, which destroyed prehistoric civilization and all records of it and plunged our ancestors into the Stone Age. The author was Ignatius Donnelly, a former politician, who is better known for his related work, *Atlantis, the Antediluvian World* (1882).

35. C. R. Longwell, The 1950 silly season, *Science* 113 (1951): 418; E. Larrabee, Two spheres collide, *Science* 118 (1953): 167; R. Gillette, Velikovsky: AAAS forum for a mild collision, *Science* 183 (1974): 1059–1062.

36. C. Lévi-Strauss, *The Savage Mind* (Chicago: University of Chicago Press, 1962). This, indeed, is such a powerful image that it was adopted as a metaphor for evolutionary biology by the French molecular biologist François Jacob, who famously argued that evolution is like a "tinkerer" rather than like an engineer. F. Jacob, Evolution and tinkering, *Science* 196 (1977): 1161–1166.

37. A. Aveni, *Conversing with the Planets: How Science and Myth Invented the Cosmos* (New York: Times Books, 1992).

38. All right, if you must know, it was D. Morris, *The Naked Ape: A Zoologist's Study of the Human Animal* (New York: Dell Publishing, 1967). Now get back to the text.

39. P. Duesberg, Retroviruses as carcinogens and pathogens: Expectations and reality, *Cancer Research* 47 (1987): 199–220; P. Duesberg, HIV is not the cause of AIDS, *Science* 241 (1988): 514–517; P. Duesberg, Human immunodeficiency virus and acquired immunodeficiency syndrome: Correlation but not causation, *Proceedings of the National Academy of Sciences USA* 86 (1989): 755–764; P. Duesberg, AIDS acquired by drug consumption and other noncontagious risk factors, *Pharmacological Therapeutics* 55 (1992): 201–277.

40. M. Ascher, H. W. Sheppard, W. Winkelstein Jr., and E. Vittinghoff, Does drug use cause AIDS? *Nature* 362 (1993): 103–104; J. Maddox, Has Duesberg a right of reply? *Nature* 363 (1993): 109; L. Chieco-Bianchi and G. B. Rossi, Duesberg: Rights and wrongs, *Nature* 364 (1993): 96; Anonymous, New-style abuse of press freedom, *Nature* 366 (1993): 493–494; J. Cohen, The Duesberg phenomenon, *Science* 266 (1994): 1642–1649; N. Hodgkinson, AIDS plagued by journalists, *Nature* 368 (1994): 387; R. Strohmann, Preface, in *Infectious AIDS: Have We Been Misled?* by Peter H. Duesberg (Berkeley, CA: North Atlantic Books, 1995), vii–xiv; J. Moore, À Duesberg, adieu! *Nature* 380 (1995): 293–294.

41. R. Lewin, *Bones of Contention: Controversies in the Search for Human Origins* (New York: Simon and Schuster, 1987); A. Walker and P. Shipman, *The Wisdom of Bones: In Search of Human Origins* (New York: Alfred A. Knopf, 1996).

42. E. Morgan, *The Descent of Woman* (New York: Stein and Day, 1972); J. Lang-

don, Umbrella hypotheses and parsimony in human evolution: A critique of the aquatic ape hypothesis, *Journal of Human Evolution* 33 (1997): 479–494.

43. Beats me. I tend to think we evolved on land, and we began to sweat because it became increasingly difficult to pant as the structure of the mouth was remodeled by the evolution of language. The efficiency of sweating for heat dissipation would require reducing the body hair.

SEVEN. SCIENTIFIC MISCONDUCT

1. W. Broad and N. Wade, *Betrayers of the Truth* (New York: Simon and Schuster, 1982); A. Kohn, *False Prophets* (New York: Basil Blackwell, 1986).

2. D. Hull, A mechanism and its metaphysics: An evolutionary account of the social and conceptual development of science, *Biology and Philosophy* 3 (1988): 123–155.

3. P. Gagneux, D. Woodruff, and C. Boesch, Furtive mating in female chimpanzees, *Nature* 387 (1997): 358–359; J. Constable, M. Ashley, J. Goodall, and A. Pusey, Noninvasive paternity assignment in Gombe chimpanzees, *Molecular Ecology* 10 (2001): 1279–1300; L. Vigilant, M. Hofreiter, H. Siedel, and C. Boesch, Paternity and relatedness in wild chimpanzee communities, *Proceedings of the National Academy of Sciences USA* 98 (2001): 12890–12895; P. Gagneux, D. Woodruff, and C. Boesch, Retraction: Furtive mating in female chimpanzees, *Nature* 414 (2001): 508. For an argument in favor of ritual humiliation as a normative scientific practice, see R. A. Gitzen, The dangers of advocacy in science, *Science* 317 (2007): 748.

4. W. Stewart and N. Feder, The integrity of the scientific literature, *Nature* 325 (1987): 207–214.

5. S. Shapin, *A Social History of Truth: Civility and Science in Seventeenth-Century England* (Chicago: University of Chicago Press, 1994).

6. M. E. Jahn and D. J. Woolf, *The Lying Stones of Dr. Beringer* (Berkeley: University of California Press, 1963); S. Gould, The lying stones of Wurzburg and Marrakech, *Natural History* 107 (1998): 16–21.

7. C. Babbage, The decline of science in England [1830], *Nature* 340 (1989): 499–502.

8. S. Williams, *Fantastic Archaeology: The Wild Side of North American Prehistory* (Philadelphia: University of Pennsylvania Press, 1991).

9. W. K. Gregory, The Dawn Man of Piltdown, England, *American Museum Journal* 14 (1914): 189–200.

10. F. Spencer, *Piltdown: A Scientific Forgery* (New York: Oxford University Press, 1990); F. Spencer, *The Piltdown Papers, 1908–1955* (New York: Oxford University Press, 1990).

11. Ultimately these positions were reconciled by modern capitalism, which takes class divisions for granted but holds their boundaries to be more permeable and not predicated on innate differences among people. One still encounters the older, polar positions periodically, and other possibilities can also be imagined.

12. The alternative, that acquired characteristics can be inherited, is often associated with the evolutionary ideas of the early nineteenth-century French biologist Lamarck but was a component as well of the quite different evolutionary ideas of the later nineteenth-century English biologist Darwin. To Lamarck, the inheritance of acquired characteristics was literally the way in which evolution occurred, arising according to the needs of the organism during its lifetime, which then produced offspring born with different (and better) properties than their parents were born with. To Darwin, the inheritance of acquired characteristics was an engine for the production of variation, upon which natural selection then acted.

13. L. A. Aronson, The case of *The Case of the Midwife Toad, Behavior Genetics* 5 (1975): 115–125; S. Gliboff, "Protoplasm . . . is soft wax in our hands": Paul Kammerer and the art of biological transformation, *Endeavour* 29 (2005): 162–167; S. Gliboff, The case of Paul Kammerer: Evolution and experimentation in the early 20th century, *Journal of the History of Biology* 39 (2006): 525–563.

14. This point was made by Stephen Jay Gould in his critique of a 1911 book ostensibly demonstrating the inheritance of feeblemindedness in the pseudonymous Kallikak family. See S. J. Gould, *The Mismeasure of Man* (New York: W. W. Norton, 1981).

15. R. Goldschmidt, Science and politics, *Science* 109 (1949): 219–227.

16. It is possible that Kammerer actually got these results through subtle, but intense, regimes of selection, of the sort that would later be called "genetic assimilation" by the geneticist C. H. Waddington.

17. Anonymous, Scientist tells of success where Darwin met failure. Eyes developed in newts. Demonstrates acquired qualities may be inherited. Austrian savant's laurels. Evolution would be speeded up if best characteristics could be transmitted, *New York Times*, June 3, 1923.

18. Ibid.

19. Anonymous, Biologist predicts critics' conversion. Dr. Kammerer not perturbed by skepticism of American scientists. Illness prevents lecture. He is unable to appear in Brooklyn and will speak first on Dec. 17, *New York Times*, December 2, 1923; L. A. Jones, Would direct evolution, *New York Times*, December 2, 1923.

20. P. Kammerer, Paul Kammerer's letter to the Moscow Academy, *Science* 64 (1926): 493–494.

21. The phrase is the title of Arthur Koestler's famous 1971 book, which is unfortunately more of a polemical novel than historiographic analysis.

22. M. LaFollette, *Stealing into Print: Fraud, Plagiarism, and Misconduct in Scientific Publishing* (Berkeley: University of California Press, 1996).

23. E. Rossiter, Reflections of a whistle-blower, *Nature* 357 (1992): 434–436; C. MacIlwain, Whistleblowers face blast of hostility, *Nature* 385 (1997): 669; E. Check, Scientific misconduct: Sitting in judgement, *Nature* 419 (2002): 332–333.

24. R. Lewin, DNA conflict continues, *Science* 241 (1998): 1756–1759; see below for more on this case. For a contrary opinion, see D. Koshland, The price of progress, *Science* 241 (1988): 637.

25. Anonymous, Germany bars a Soviet film, *New York Times*, February 17, 1929.

26. C. Zirkle, Citation of fraudulent data, *Science* 120 (1954): 189–190; J. Campanario, Fraud: Retracted articles are still being cited, *Nature* 408 (2000): 288.

27. R. F. Service, Bell Labs fires star physicist found guilty of forging data, *Science* 298 (2002): 30–31.

28. The quotation explaining the action is from Nicole Suciu-Foca, editor in chief of *Human Immunology*, in *Nature Genetics* 30 (2002): 140. A. Arnaiz-Villena, N. Elaiwa, C. Silvera, A. Rostom, J. Moscoso, E. Gómez-Casado, L. Allende, P. Varela, and J. Martínez-Laso, The origin of Palestinians and their genetic relatedness with other Mediterranean populations, *Human Immunology* 62 (2001): 889–900; E. Klarreich, Genetics paper erased from journal over political content, *Nature* 414 (2001): 382; N. Risch, A. Piazza, and L. Cavalli-Sforza, Dropped genetics paper lacked scientific merit, *Nature* 415 (2002): 115; A. Arnaiz-Villena, E. Gomez-Casado, and J. Martinez-Laso, Single-locus studies, *Nature* 416 (2002): 677; S. Krimsky, For the record, *Nature Genetics* 30 (2002): 139; K. Shashok, Pitfalls of editorial miscommunication, *British Medical Journal* 326 (2003): 1262–1264. Access the paper at http://rense.com/general48/Palestinians.pdf; or http://giwersworld.org/hidden/jewish-gene-paper.pdf; http://kinoko.c.u-tokyo.ac.jp/~duraid/stolen_science/The_Origin_of_Palestinians_and_Their_Genetic_Relatedness_With_Other_Mediterranean_Populations.pdf (accessed September 3, 2007).

29. J. Mervis, Don't steal this book, *Nature* 359 (1992): 787.

30. J. E. Brody, Lab discovery may aid transplants, *New York Times*, March 31, 1973; L. Edson, A secret weapon called immunology, *New York Times Magazine*, February 17, 1974; J. E. Brody, Charge of false research data stirs cancer scientists at Sloan-Kettering, *New York Times*, April 18, 1974; J. E. Brody, Inquiry at cancer center finds fraud in research, *New York Times*, May 25, 1974; J. E. Brody, Scientist denies cancer research fraud, *New York Times*, May 29, 1974; B. Culliton, The Sloan-Kettering affair: A story without a hero, *Science* 184 (1974): 644–650; B. Culliton, The Sloan-Kettering affair (II): An uneasy resolution, *Science* 184 (1974):

1154–1157; J. Hixson, *The Patchwork Mouse: Politics and Intrigue in the Campaign to Conquer Cancer* (New York: Anchor Press, 1976); P. Basu, Where are they now? *Nature Medicine* 12 (2006): 492–493.

31. W. Broad, Imbroglio at Yale (I): Emergence of a fraud, *Science* 210 (1980): 38–41; W. Broad, Imbroglio at Yale (II): A top job lost, *Science* 210 (1980): 171–173; M. Hunt, A fraud that shook the world of science, *New York Times*, November 1, 1981.

32. E. Racker, A view of misconduct in science, *Nature* 339 (1989): 91–93.

33. W. Broad, Congress told fraud issue "exaggerated," *Science* 212 (1981): 421; D. Koshland, Fraud in science, *Science* 235 (1987): 141; D. Goodstein, Scientific fraud, *American Scholar* 60 (1991): 505–515.

34. E. Marshall, How prevalent is fraud? That's a million dollar question, *Science* 290 (2000): 1662–1663; M. Benvenuto, Ethical behaviour as a stakes game, *Science* 291 (2000): 2316.

35. D.P. Hamilton, In the trenches, doubts about scientific integrity, *Science* 255 (1991): 1636; J. Swazey, M. Anderson, and K. Louis, Ethical problems in academic research, *American Scientist* 81 (1993): 542–553.

36. B.C. Martinson, M.S. Anderson, and R. deVries, Scientists behaving badly, *Nature* 435 (2005): 737–738.

37. *On Being a Scientist* (Washington, DC: National Academy Press, 1995).

38. J. Sarasohn, *Science on Trial: The Whistle Blower, the Accused, and the Nobel Laureate* (New York: St. Martin's Press, 1993); D. Kevles, *The Baltimore Case: A Trial of Politics, Science, and Character* (New York: W.W. Norton, 2000); H.F. Judson, *The Great Betrayal* (Orlando, FL: Harcourt, 2004).

39. M. O'Toole, The whistle-blower and the train wreck, *New York Times*, April 12, 1991.

40. C. Holden, Whistle-blowers air cases at House hearings, *Science* 240 (1988): 386–387; H. Wortis, B. Huber, and R. Woodland, Fraud allegations, *Science* 240 (1988): 968

41. B. Culliton, Dingell v. Baltimore, *Science* 244 (1989): 412–414; B. J. Culliton, The Dingell probe finally goes public, *Science* 244 (1989): 643–646; Kevles, *The Baltimore Case;* Sarasohn, *Science on Trial;* Judson, *The Great Betrayal.*

42. S. Gould, Judging the perils of official hostility to scientific error, *New York Times*, July 30, 1989.

43. D. P. Hamilton, NIH finds fraud in *Cell* paper, *Science* 251 (1991): 1552–1554.

44. P. Doty, Responsibility and Weaver et al., *Nature* 352 (1991): 183–184.

45. Anonymous, Baltimore defeat a defeat for research, *Nature* 354 (1991): 419–420.

46. Holden, Whistle-blowers air cases at House hearings; Wortis, Huber, and Woodland, Fraud allegations; J. Foreman, Baltimore speaks out on disputed

study in letter sent to colleagues around the nation, he calls for protection against "threats" to scientific freedom, *Boston Globe,* May 23, 1988; Culliton, Dingell v. Baltimore; Gould, Judging the perils of official hostility; Hamilton, NIH finds fraud in *Cell* paper.

47. P. Doty, Baltimore's unanswered questions, *Nature* 353 (1991): 495.

48. S.L. Titus, J.A. Wells, and L.J. Rhoades, Repairing research integrity, *Nature* 453 (2008): 980–982.

49. A. Jensen, Kinship correlations reported by Sir Cyril Burt, *Behavioral Genetics* 4 (1974): 1–28; L. Kamin, *The Science and Politics of I.Q.* (Potomac, MD: Erlbaum, 1974); N. Wade, IQ and heredity: Suspicion of fraud beclouds classic experiment, *Science* 194 (1976): 916–919; L. Kamin, Burt's IQ data, *Science* 195 (1977): 246–248; D.D. Dorfman, The Cyril Burt question: New findings, *Science* 201 (1978): 1177–1186; D.D. Dorfman, Burt's tables, *Science* 204 (1979): 246–254; O. Gillie, Burt's missing ladies, *Science* 204 (1979): 1035–1039; L. Hearnshaw, *Cyril Burt, Psychologist* (Ithaca, NY: Cornell University Press, 1979); N. Hawkes, Tracing Burt's descent to scientific fraud, *Science* 205 (1979): 673–675; R.E. Fancher, The Burt case: Another foray, *Science* 253 (1991): 1565–1566.

50. D.C. Gajdusek, Scientific responsibility, in *Proceedings of the Second International Bioethics Seminar in Fukui, Japan,* ed. N. Fujiki and D. Macer, 205–210 (Christchurch, NZ: Eubios Ethics Institute, 1992); N. Molotsky, Nobel scientist pleads guilty to abusing boy, *New York Times,* February 19, 1997; J. Cohen, The culture of credit, *Science* 268 (1995): 1706–1711; P.Y. Hong, Scientist gets 14 years for molesting girl, *Los Angeles Times,* February 3, 2007; D. Overbye, Crossed by the stars they reach for, *New York Times,* February 13, 2007.

51. A. Lubasch, N.Y.U. professor gets five years in illicit manufacturing of drugs, *New York Times,* November 14, 1980; B. Lambert, John Buettner-Janusch, 67, dies; N.Y.U. professor poisoned candy, *New York Times,* July 4, 1992.

52. "Their methods were logical, and made very little difference to inferences from the complete data": J.A.W. Kirsch and C. Krajewski, Reviewing misconduct? *American Scientist* 81 (1993): 410. "[Without the alterations] it is virtually certain that Sibley and Ahlquist would have concluded that *Homo, Pan,* and *Gorilla* form a trichotomy": C.G. Sibley, J.A. Comstock, and J.E. Ahlquist, DNA hybridization evidence of hominoid phylogeny: A reanalysis of the data, *Journal of Molecular Evolution* 30 (1990): 225.

53. F. Gill and F. Sheldon, The birds reclassified, *Science* 252 (1991): 1003–1005.

54. C. Gammon, The case of the missing eggs, *Sports Illustrated,* June 24, 1974; C.G. Sibley and J.E. Ahlquist, The phylogeny of the hominoid primates, as indicated by DNA-DNA hybridization, *Journal of Molecular Evolution* 20 (1984): 2–15; C.G. Sibley and J.E. Ahlquist, DNA hybridization evidence of hominoid phylogeny: Evidence from an expanded data set, *Journal of Molecular Evolution* 26 (1987):

99–121; R. Lewin, Conflict over DNA clock results, *Science* 241 (1988): 1598–1600; J. Marks, Relationships of humans to chimps and gorillas, *Nature* 334 (1988): 656; J. Marks, C.W. Schmid, and V.M. Sarich, DNA hybridization as a guide to phylogeny: Relations of the Hominoidea, *Journal of Human Evolution* 17 (1988): 769–786; V.M. Sarich, C.W. Schmid, and J. Marks, DNA hybridization as a guide to phylogeny: A critical analysis, *Cladistics* 5 (1988): 3–32; J. Horgan, Time bomb: War breaks out in the field of evolutionary biology, *Scientific American* 260, no. 3 (1989): 24–25; C.G. Sibley, J. A. Comstock, and J.E. Ahlquist, DNA hybridization evidence of hominoid phylogeny: A reanalysis of the data, *Journal of Molecular Evolution* 30 (1990): 202–236; A.M. Highley, DNA hybridization, *Yale Scientific* 65 (1991): 6–10; J. Marks, What's old and new in molecular phylogenetics, *American Journal of Physical Anthropology* 85 (1991): 207–219; Gill and Sheldon, The birds reclassified.

55. J. Marks, Scientific misconduct: Where "Just say no" fails, *American Scientist* 81 (1993): 380–382.

56. C. Sibley and J. Ahlquist, Reviewing misconduct? *American Scientist* 81 (1993): 407–408.

EIGHT. THE RISE AND FALL OF COLONIAL SCIENCE

1. R. Silverberg, *Mound Builders of Ancient America: The Archaeology of a Myth* (Greenwich, CT: New York Graphic Society, 1968); B. Trigger, Archaeology and the image of the American Indian, *American Antiquity* 45 (1980): 662–676; J. Sabloff and G.R. Willey, *A History of American Archaeology* (San Francisco: W.H. Freeman, 1980); D. Blakeslee, John Rowzee Peyton and the myth of the Mound Builders, *American Antiquity* 52 (1987): 784–792; R. McGuire, Archeology and the first Americans, *American Anthropologist* 94 (1992): 816–836.

2. J. T. Bent, On the finds at the Great Zimbabwe ruins (with a view to elucidating the origin of the race that built them), *Journal of the Anthropological Institute of Great Britain and Ireland* 22 (1893): 123–136; D. G. Brinton, Current notes on anthropology, VIII, *Science* 19 (1892): 343.

3. R. N. Hall, in The Rhodesia ruins: Their probable origin and significance: Discussion, by Arthur Evans et al., *Geographical Journal* 27 (1906): 336–347; G. Caton-Thompson, Recent excavations at Zimbabwe and other ruins in Rhodesia, *Journal of the Royal African Society* 29 (1930): 132–138.

4. Bent, On the finds at the Great Zimbabwe ruins; D. Randall-MacIver, The Rhodesia ruins: Their probable origin and significance, *Geographical Journal* 27 (1906): 325–347; Caton-Thompson, Recent excavations at Zimbabwe and other ruins in Rhodesia; D. Tangri, Popular fiction and the Zimbabwe controversy, *History in Africa* 17 (1990): 293–304; M. Hall, Heads and tales, *Representations* 54 (1996): 104–123.

5. C.B. Davenport, *Heredity in Relation to Eugenics* (New York: Henry Holt, 1911), 214.

6. L. Van Gelder, C.P. Snow says Jews' success could be genetic superiority, *New York Times*, April 1, 1969.

7. C. Quigley, Review of *The Evolution of Man and Society* by C.D. Darlington, *American Anthropologist* 73 (1971): 434–439; P. Medawar, The volubility of DNA, *New York Review of Books* 16, no. 9 (1971).

8. N. Wade, *Before the Dawn: Recovering the Lost History of Our Ancestors* (New York: Penguin, 2006); K. Weiss and A. Buchanan, In your own image: Care must be taken when looking for natural selection to explain the evolution of human behaviour, *Nature* 441 (2006): 813–814; R. Cann, Human evolution: Silverbacks and their satellites, *Science* 313 (2006): 174; H.A. Orr, Talking genes, *New York Review of Books* 53, no. 14 (2006).

9. A. Crosby, *Ecological Imperialism: The Biological Expansion of Europe, 900–1900* (New York: Cambridge University Press, 1986).

10. L.A. White, *The Science of Culture* (New York: Farrar, Straus, 1949); M. Sahlins and E. Service, *Evolution and Culture* (Ann Arbor: University of Michigan Press, 1960); W.H. McNeill, History upside down, *New York Review of Books* 44, no. 8 (1997); M.M.J. Fischer, Culture and cultural analysis as experimental systems, *Cultural Anthropology* 22 (2007): 1–65.

11. G.A. Dorsey, The Department of Anthropology of the Field Columbian Museum: A review of six years, *American Anthropologist* 2 (1900): 247–265.

12. M. S. Lindee, *Suffering Made Real: American Science and the Survivors at Hiroshima* (Chicago: University of Chicago Press, 1994).

13. G. Basalla, The spread of Western science, *Science* 156 (1967): 611–622; R. MacLeod, *Nature and Empire: Science and the Colonial Enterprise* (Chicago: University of Chicago Press, 2001); L. Schiebinger, *Plants and Empire: Colonial Bioprospecting in the Atlantic World* (Cambridge, MA: Harvard University Press, 2004); K. Philip, *Civilizing Natures: Race, Resources, and Modernity in Colonial South India* (New Brunswick, NJ: Rutgers University Press, 2004); M. Harrison, Science and the British empire, *Isis* 9 (2005): 56–63.

14. There are many derivative citations of this quote, but it is difficult to find a primary citation. But even if Rhodes didn't exactly commit this to paper, he certainly thought it.

15. G.D. Stone, Both sides now: Fallacies in the genetic-modification wars, implications for developing countries, and anthropological perspectives, *Current Anthropology* 43, no. 4 (2002): 611–630.

16. H. Kuklick, *The Savage Within: A History of British Anthropology, 1885–1945* (New York: Cambridge University Press, 1991); P. Bourdieu, Colonialism and ethnography, *Anthropology Today* 19 (2003): 13–18; A. Kuper, Alternative histories of British social anthropology, *Social Anthropology* 13 (2005): 47–64; D. Segal and

S. Yanagisako, eds., *Unwrapping the Sacred Bundle: Reflections on the Disciplining of Anthropology* (Durham, NC: Duke University Press, 2005).

17. Anonymous, Explorer Verner home with African curios. Brings Mr. Otabenga, a dwarf, to act as his valet. A man in a boy's body. The pygmy thinks Peary meteorite a strange god—made blasé by marvels he has seen, *New York Times*, September 2, 1906; Anonymous, Man and monkey show disapproved by clergy. The Rev. Dr. MacArthur thinks the exhibition degrading. Colored ministers to act. The pygmy has an orang-outang as a companion now and their antics delight the Bronx crowds, *New York Times*, September 10, 1906; Anonymous, Negro ministers act to free the pygmy. Will ask the Mayor to have him taken from the monkey cage. Committee visits the zoo. Public exhibitions of the dwarf discontinued, but will be resumed, Mr. Hornaday says, *New York Times*, September 11, 1906; Anonymous, The Mayor won't help to free caged pygmy. He refers Negro ministers to the Zoological Society. Crowd annoys the dwarf. Failing to get action from other sources, the committee will ask the courts to interfere, *New York Times*, September 12, 1906; Anonymous, African pygmy's fate is still undecided. Director Hornaday of the Bronx Park throws up his hands. Asylum doesn't take him. Benga meanwhile laughs and plays with a ball and mouth organ at the same time, *New York Times*, September 18, 1906; Anonymous, Colored orphan home gets the pygmy. He has a room to himself and may smoke if he likes. To be educated if possible. When he returns to the Congo he may then help to civilize his people, *New York Times*, September 29, 1906; Anonymous, Ota Benga, pygmy, tired of America. The strange little African finally ended life at Lynchburg, Va. Once at the Bronx Zoo. His American sponsor found him shrewd and courageous—wanted to be educated, *New York Times*, September 11, 1916; P. V. Bradford and H. Blume, *Ota Benga: The Pygmy in the Zoo* (New York: Delta, 1993).

18. C. Anstötz, Profoundly intellectually disabled humans and the great apes: A comparison, in *The Great Ape Project: Equality beyond Humanity*, ed. P. Cavalieri and P. Singer, 158–172 (New York: St. Martin's Press, 1993); N. Groce and J. Marks, The Great Ape Project and disability rights: Ominous undercurrents of eugenics in action, *American Anthropologist* 102 (2000): 818–822.

19. P. Rabinow, *Essays on the Anthropology of Reason* (Princeton, NJ: Princeton University Press, 1996), 129–152; R. Skloot, Taking the least of you, *New York Times Magazine*, April 16, 2007.

20. S. Cole, *Leakey's Luck: The Life of Louis Seymour Bazett Leakey, 1903–1972* (London: Collins, 1975); V. Morell, *Ancestral Passions: The Leakey Family and the Quest for Humankind's Beginnings* (New York: Simon and Schuster, 1995); L. S. B. Leakey, New finds at Olduvai Gorge, *Nature* 189 (1961): 650.

21. L. S. B. Leakey and M. D. Leakey, Recent discoveries of fossil hominids in Tanganyika: At Olduvai and near Lake Natron, *Nature* 202 (1964): 5–7.

22. H. Landecker, Immortality, in vitro: A history of the HeLa cell line, in *Biotechnology and Culture: Bodies, Anxieties, Ethics,* ed. P. Brodwin, 53–72 (Bloomington: Indiana University Press, 2000); R. Skloot, Henrietta's dance, *Johns Hopkins Magazine,* April 2000; R. Skloot, *The Immortal Life of Henrietta Lacks* (New York: Crown, in press).

23. K.B. Strier, An American primatologist abroad in Brazil, in *Primate Encounters,* ed. S.C. Strum and L.M. Fedigan, 194–207 (Chicago: University of Chicago Press, 2000); K. Strier, J. Boubli, F. Pontual, and S. Mendes, Human dimensions of northern muriqui conservation efforts, *Ecological and Environmental Anthropology* 2 (2007): 44–53.

24. L. Schiebinger, *Nature's Body* (Boston: Beacon, 1993); P. Tobias, Saartje Baartman: Her life, her remains, and the negotiations for their repatriation from France to South Africa, *South African Journal of Science* 98 (2002): 107–110.

25. Anonymous, Returned from the Arctic. Meteorite from the Hope to be unloaded at Brooklyn—Six Eskimos for Boston, *New York Times,* September 27, 1897; Anonymous, Back from the far north. Steamer Hope, with the big Greenland meteorite aboard, moored in Brooklyn. Eskimos and curiosities. Lieut. Peary brings relics from the Greely camp and denies the reported finding of evidences of cannibalism, *New York Times,* October 1, 1897; Anonymous, Too warm for Eskimos. Unaccustomed to a temperate climate, they are suffering with a hitherto unknown complaint. Their spirits unaffected. At an informal reception held at the Museum of Natural History they entertain their guests and make love to the women, *New York Times,* October 11, 1897.

26. I. Jacknis, The First Boasian: Alfred Kroeber and Franz Boas, 1896–1905, *American Anthropologist* 104 (2002): 520–532.

27. Anonymous, Kushan, the Eskimo, dead. Our climate too much for him—Natural History Museum wants his skeleton, *New York Times,* February 19, 1898; A. Hrdlička, An Eskimo brain, *American Anthropologist* 3 (1901): 454–500.

28. Anonymous, Eskimo lad very ill. Last survivor of six Eskimos brought from Greenland by Peary, *New York Times,* March 1, 1904; Anonymous, Peary's athletic protégé. Mene Peary Wallace, an Eskimo, picked to win skating honors, *New York Times,* February 12, 1905; Anonymous, "Mene" to be an usher. Eskimo brought here by Peary employed by Madison Square Theatre, *New York Times,* October 28, 1907.

29. Anonymous, Peary Eskimo Mene dying. He came here with the explorer ten years ago—has pneumonia, *New York Times,* November 18, 1908.

30. Anonymous, Mene gone to balk Peary? Eskimo boy's letter to a friend says he is on his way north, *New York Times,* April 14, 1909; Anonymous, Latest bulletin from Mene. New letter says Eskimo boy is thinking of suicide, *New York Times,* May 23, 1909; Anonymous, Mene Wallace going home. Eskimo will be sent back on the Jeanie if he'll promise to stay, *New York Times,* July 9, 1909.

31. K. Harper, *Give Me My Father's Body: The Life of Minik, the New York Eskimo* (South Royalton, VT: Steerforth Press, 2000).

32. M.T. Kaufman, About New York: A museum's Eskimo skeletons and its own, *New York Times,* August 21, 1993; D. Smith, An Eskimo boy and injustice in old New York: A campaigning writer indicts an explorer and a museum, *New York Times,* March 15, 2000.

33. M. Curtius, Ishi: Group seeks to rebury tribe's last survivor in homeland, *Los Angeles Times,* June 8, 1997.

34. Anonymous, Find a rare aborigine. Scientists obtain valuable tribal lore from Southern Yahi Indian, *New York Times,* September 11, 1911; Anonymous, Hunting big game with bow and arrow. The ancient method demands more of the sportsman than does the rifle [Review of *Hunting with the Bow and Arrow,* by Saxton Pope], *New York Times,* January 10, 1926; T. Kroeber, *Ishi in Two Worlds: A Biography of the Last Wild Indian in North America* (New York: John Wiley, 1961); N. Scheper-Hughes, Ishi's brain, Ishi's ashes: Anthropology and genocide, *Anthropology Today* 17 (2001): 12–18; O. Starn, *Ishi's Brain: In Search of America's Last "Wild" Indian* (New York: W.W. Norton, 2004).

35. J. Heller, Syphilis victims in US study went untreated for 40 years, *New York Times,* July 26, 1972; C. Wooten, Survivor of '32 syphilis study recalls a diagnosis, *New York Times,* July 27, 1972; Anonymous, Aide questioned syphilis study, *New York Times,* August 9, 1972; A. Mitchell, Clinton regrets "clearly racist" U.S. study, *New York Times,* May 17, 1997; J. Jones, *Bad Blood: The Tuskegee Syphilis Experiment* (New York: Free Press, 1993).

36. P. Skotnes, *Miscast: Negotiating the Presence of the Bushmen* (Cape Town, South Africa: University of Cape Town Press, 1996).

37. N. Wolf, A woman's place, *New York Times,* May 31, 1992; D. Cavett, Women alone can't make a just society: "Graduation from Hell," *New York Times,* June 12, 1992; G. Hersey, A secret lies hidden in Vassar and Yale nude "posture photos," *New York Times,* July 3, 1992; R. Rosenbaum, The great Ivy League nude posture photo scandal, *New York Times Magazine,* January 15, 1995.

38. George Hersey died while I was writing this chapter. I will cherish my memories of him.

39. L.L. Cavalli-Sforza, A.C. Wilson, C.R. Cantor, R.M. Cook-Deegan, and M.-C. King, Call for a worldwide survey of human genetic diversity: A vanishing opportunity for the Human Genome Project, *Genomics* 11 (1991): 490–491; National Research Council, *Evaluating Human Genetic Diversity* (Washington, DC: National Academy Press, 1997); C. MacIlwain, Diversity project "does not merit federal funding," *Nature* 389 (1997): 774; H.T. Greely, Legal, ethical, and social issues in human genome research, *Annual Review of Anthropology* 27 (1998): 473–502; J. Reardon, *Race to the Finish: Identity and Governance in an Age of Genomics* (Princeton, NJ: Princeton University Press, 2004).

40. A. Harmon, DNA gatherers hit a snag: The tribes don't trust them, *New York Times,* December 10, 2006.

41. www.nationalgeographicexpeditions.com/536.html, accessed October 2007. But the tour is over, and they have taken the page down. I hope everybody had fun!

NINE. RACIAL AND GENDERED SCIENCE

1. L. B. Andrews, *The Clone Age: Adventures in the New World of Reproductive Technology* (New York: Henry Holt, 1999).

2. L. Jaroff, The gene hunt, *Time,* March 20, 1989, 62–67.

3. D. Nelkin and M. S. Lindee, *The DNA Mystique: The Gene as Cultural Icon* (New York: Freeman, 1995); M. S. Lindee, Watson's world, *Science* 300 (2003): 432–434.

4. T. Abate, Nobel winner's theories raise uproar in Berkeley: Geneticist's views strike many as racist, sexist, *San Francisco Chronicle,* November 13, 2000; E. O. Wilson, *Naturalist* (New York: Island Press, 1994).

5. C. Hunt-Grubbe, The elementary DNA of Dr Watson, *Sunday Times* (London), October 14, 2007.

6. C. Milmo, Fury at DNA pioneer's theory: Africans are less intelligent than Westerners, *Independent* (London), October 17, 2007.

7. D. Plotz, *The Genius Factory: The Curious History of the Nobel Prize Sperm Bank* (New York: Random House, 2005); J. N. Shurkin, *Broken Genius: The Rise and Fall of William Shockley, Creator of the Electronic Age* (New York: Macmillan, 2006).

8. D. J. Kevles, Genetics, race, and IQ: Historical reflections from Binet to *The Bell Curve, Contention* 5 (1995): 3–18; F. Samelson, On the uses of history: The case of *The Bell Curve, Journal of the History of the Behavioral Sciences* 33 (1997): 129–133.

9. C. S. Fischer, M. Hout, M. S. Jankowski, S. R. Lucas, A. Swidler, and K. Voss, *Inequality by Design: Cracking the Bell Curve Myth* (Princeton, NJ: Princeton University Press, 1996); J. L. Kincheloe, S. R. Steinberg, and A. D. Gresson III, eds., *Measured Lies: The Bell Curve Examined* (New York: St. Martin's Press, 1996); J. Ogbu, Cultural amplifiers of intelligence: IQ and minority status in cross-cultural perspective, in *Race and Intelligence: Separating Science from Myth,* ed. J. Fish, 241–278 (Mahwah, NJ: Lawrence Erlbaum, 2002); J. Marks, Anthropology and *The Bell Curve,* in *Why America's Top Pundits Are Wrong: Anthropologists Talk Back,* ed. C. Besteman and H. Gusterson, 206–227 (Berkeley: University of California Press, 2005).

10. R. C. Lewontin, Race and intelligence, *Bulletin of the Atomic Scientists* 26

(1970): 2–8; J. Marks, *Human Biodiversity: Genes, Race, and History* (New York: Aldine de Gruyter, 1995); D. Wahlsten, The malleability of intelligence is not constrained by heritability, in *Intelligence, Genes, and Success: Scientists Respond to* The Bell Curve, ed. B. Devlin, S.E. Fienberg, D.P. Resnick, and K. Roeder, 71–87 (New York: Copernicus, 1997).

11. M. Henderson, Gay worms get down and dirty with their mates, *Times* (London), October 26, 2007.

12. M. Fortun, The Human Genome Project: Past, present, and future anterior, *Boston Studies in the Philosophy of Science* 228 (2001): 339–362; N. Brown, Hope against hype: Accountability in biopasts, presents and futures, *Science Studies* 16 (2003): 3–21; M. Fortun, For an ethics of promising, or: a few kind words about James Watson, *New Genetics and Society* 24 (2005): 157–173.

13. N.A. Holtzman, Are genetic tests adequately regulated? *Science* 286 (1999): 409.

14. Anonymous, Eugenics will end cancer, woman scientist declares. Dr. Slye of Chicago University reports results of experiments with mice—thinks right marrying will uproot disease, *New York Evening Post*, August 3, 1929.

15. N. Wade, Genome of DNA discoverer is deciphered, *New York Times*, June 1, 2007; J. Shreeve, *The Genome War: How Craig Venter Tried to Capture the Code of Life and Save the World* (New York: Alfred A. Knopf, 2004).

16. A.E. Wiggam, *The Fruit of the Family Tree* (Indianapolis, IN: Bobbs-Merrill, 1924), 1.

17. Nelkin and Lindee, *DNA Mystique*; P. Buerton, R. Falk, and H.-J. Rheinberger, eds., *The Concept of the Gene in Development and Evolution* (New York: Cambridge University Press, 2000); L. Moss, *What Genes Can't Do* (Cambridge, MA: MIT Press, 2003).

18. R. Hubbard and E. Wald, *Exploding the Gene Myth: How Genetic Information Is Produced and Manipulated by Scientists, Physicians, Employers, Insurance Companies, Educators, and Law Enforcers* (Boston: Beacon Press, 1993).

19. E.O. Manoiloff, Discernment of human races by blood: Particularly of Russians from Jews, *American Journal of Physical Anthropology* 10 (1927): 11–21; A.T. Poliakowa, Manoiloff's "race" reaction and its application to the determination of paternity, *American Journal of Physical Anthropology* 10 (1927): 23–29; N.P. Naidoo, G. Štrkalj, and T.J.M. Daly, The alchemy of human variation: Race, ethnicity and Manoiloff's blood reaction, *Anthropological Review* 70 (2007): 37–43.

20. W.Z. Ripley, *The Races of Europe* (New York: D. Appleton, 1899); C.S. Coon, *The Races of Europe* (New York: Macmillan, 1939).

21. W.C. Boyd, Genetics and the human race, *Science* 140 (1963): 1057–1065.

22. F.P. Thieme, The population as a unit of study, *American Anthropologist* 54 (1952): 504–509; J.S. Weiner, Physical anthropology: An appraisal, *American Scientist* 45 (1957): 79–87; F.S. Hulse, Race as an evolutionary episode, *American*

Anthropologist 64 (1962): 929–945; F.E. Johnston, The population approach to human variation, *Annals of the New York Academy of Sciences* 134 (1966): 507–515.

23. M.F.A. Montagu, The concept of race in the light of genetics, *Journal of Heredity* 23 (1941): 243–427; F.B. Livingstone, On the non-existence of human races, *Current Anthropology* 3 (1962): 279–281; R.C. Lewontin, The apportionment of human diversity, *Evolutionary Biology* 6 (1972): 381–398; G. Barbujani, A. Magagni, E. Minch, and L.L. Cavalli-Sforza, An apportionment of human DNA diversity, *Proceedings of the National Academy of Sciences USA* 94 (1997): 4516–4519; J. Marks, Long shadow of Linnaeus's human taxonomy, *Nature* 447 (2007): 28; J. Marks, Race: Past, present, and future, in *Revisiting Race in a Genomic Age,* ed. B.A. Koenig, S.S.-J. Lee, and S. Richardson (Piscataway, NJ: Rutgers University Press, in press).

24. B. Campbell, The systematics of man, *Nature* 194 (1962): 225–232; J. Marks, *Human Biodiversity: Genes, Race, and History* (New York: Aldine de Gruyter, 1995).

25. C.G. Seligman, *The Races of Africa* (Oxford, UK: Oxford University Press, 1922).

26. D.A. Bolnick, Individual ancestry inference and the reification of race as a biological phenomenon, in *Revisiting Race in a Genomic Age,* ed. B.A. Koenig, S.S.-J. Lee, and S. Richardson (Piscataway, NJ: Rutgers University Press, in press).

27. D.A. Bolnick, D. Fullwiley, T. Duster, R.S. Cooper, J. Fujimura, J. Kahn, J. Kaufman, J. Marks, A. Morning, A. Nelson, P. Ossorio, J. Reardon, S. Reverby, and K. Tallbear, The science and business of genetic ancestry testing, *Science* 318 (2007): 399–400.

28. S. Swinford, DNA reunited, *Times* (London), October 28, 2007.

29. R.J. David and J.W. Collins Jr., Differing birth weights among infants of U.S.-born blacks, African-born blacks, and U.S.-born whites, *New England Journal of Medicine* 337 (1997): 1209–1214; A. Goodman, Why genes don't count (for racial differences in health), *American Journal of Public Health* 90 (2000): 1699–1701; R.S. Garcia, The misuse of race in medical diagnosis, *Chronicle of Higher Education* 49 (2003): B15; J. Kaufman and S. Hall, The slavery hypertension hypothesis: Dissemination and appeal of a modern race theory, *Epidemiology and Society* 14 (2003): 111–126; D. Martins and K. Norris, Hypertension treatment in African Americans: Physiology is less important than sociology, *Cleveland Clinic Journal of Medicine* 71 (2004): 735–743; W. Dressler, K. Oths, and C. Gravlee, Race and ethnicity in public health research: Models to explain health disparities, *Annual Review of Anthropology* 34 (2005): 231–252.

30. R.F. Service, Going from genome to pill, *Science* 308 (2005): 1858–1860; S. Saul, 2 officials quit amid slow sales of heart drug for blacks, *New York Times,* March 22, 2006; J. Kahn, Patenting race, *Nature Biotechnology* 24 (2006): 1349–1351;

K. Bibbins-Domingo and A. Fernandez, BiDil for heart failure in black patients: Implications of the U.S. Food and Drug Administration approval, *Annals of Internal Medicine* 146 (2007): 52–56.

31. S. Satel, I am a racially profiling doctor, *New York Times,* May 5, 2002.

32. V.M. Sarich and F. Miele, *Race: The Reality of Human Differences* (New York: Westview, 2004); A. Leroi, A family tree in every gene, *New York Times,* March 14, 2005; O. Judson, The subject is taboo, *New York Times,* June 28, 2006, http://judson.blogs.nytimes.com/2006/06/28/the-subject-is-taboo/; N. Wade, *Before the Dawn: Recovering the Lost History of Our Ancestors* (New York: Penguin, 2006).

33. D.C. Dennett, *Darwin's Dangerous Idea: Evolution and the Meanings of Life* (New York: Simon and Schuster, 1995); S. Pinker, *The Blank Slate: The Modern Denial of Human Nature* (New York: Viking Penguin, 2002); S. McKinnon, *Neoliberal Genetics: The Myths and Moral Tales of Evolutionary Psychology* (Chicago: Prickly Paradigm Press, 2005); R. C. Richardson, *Evolutionary Psychology as Maladpated Psychology* (Cambridge, MA: MIT Press, 2007).

34. D. Buss, *The Evolution of Desire* (New York: Viking Penguin, 1994); D. Evans and O. Zarate, *Introducing Evolutionary Psychology* (Cambridge, UK: Icon Books, 2006); R. Dunbar, L. Barrett, and J. Lycett, *Evolutionary Psychology* (Oxford, UK: Oneworld, 2007).

35. D.W. Yu and G.H. Shepard, Is beauty in the eye of the beholder? *Nature* 326 (1998): 391–392; F. Marlowe and A. Wetsman, Preferred waist-to-hip ratio and ecology, *Personality and Individual Differences* 30 (2001): 481–489.

36. M. Montoya, Bioethnic conscription: Genes, race and Mexicana/o ethnicity in diabetes research, *Cultural Anthropology* 22 (2007): 94–128.

37. M. Mead, *Male and Female* (New York: William Morrow, 1949); M. Di Leonardo, *Exotics at Home* (Chicago: University of Chicago Press, 1998).

38. S. Dillon, Harvard chief defends his talk on women, *New York Times,* January 18, 2005; C. Murray, Sex ed at Harvard, *New York Times,* January 23, 2005; O. Judson, Different but (probably) equal, *New York Times,* January 23, 2005; J. Traub, Lawrence Summers, provocateur, *New York Times,* January 23, 2005; S. Rimer, At Harvard, the bigger concern of the faculty is the president's management style, *New York Times,* January 26, 2005.

39. T. Dobzhansky, *Heredity and the Nature of Man* (New York: Harcourt, Brace and World, 1964); S. Lindee, *Moments of Truth in Genetic Medicine* (Baltimore, MD: Johns Hopkins University Press, 2005).

40. Charles Darwin to Thomas Huxley, July 17, 1865, Darwin letters, no. 4872.

41. J. Carroll, *Evolution and Literary Theory* (Columbia: University of Missouri Press, 1995); M. O'Brien and R. Lyman, *Applying Evolutionary Archaeology: A Systematic Approach* (New York: Springer, 2000); R. Nesse and G. Williams, *Why We*

Get Sick: The New Science of Darwinian Medicine (New York: Vintage Books, 1996).

42. J. P. Rushton, *Race, Evolution, and Behavior* (New Brunswick, NJ: Transaction Publishers, 1995); R. Dawkins, *The Selfish Gene* (New York: Oxford University Press, 1976); R. Thornhill and C. Palmer, *A Natural History of Rape: Biological Bases of Sexual Coercion* (Cambridge, MA: MIT Press, 2000); M. Pernick, *The Black Stork* (New York: Oxford University Press, 1996); L. Betzig, M. Mulder, and P. Turke, *Human Reproductive Behaviour: A Darwinian Perspective* (New York: Cambridge University Press, 1988); J. Diamond, *The Third Chimpanzee* (New York: HarperCollins, 1992).

43. B. Massin, From Virchow to Fischer: Physical anthropology and "modern race theories" in Wilhelmine Germany, in *Volksgeist as Method and Ethic: Essays on Boasian Ethnography and the German Anthropological Tradition*, ed. G. Stocking, 79–154 (Madison: University of Wisconsin Press, 1996).

44. T. Preuss, The discovery of cerebral diversity: An unwelcome scientific revolution, in *Evolutionary Anatomy of the Primate Cerebral Cortex*, ed. D. Falk and K. Gibson, 138–155 (Cambridge: Cambridge University Press, 2001); M. Balter, Neuroanatomy: Brain evolution studies go micro, *Science* 315 (2007): 1208.

45. C. Stanford, *Significant Others: The Ape-Human Continuum and the Quest for Human Nature* (New York: Basic Books, 2001), xi.

TEN. NATURE/CULTURE

1. T. Gieryn, *Cultural Boundaries of Science: Credibility on the Line* (Chicago: University of Chicago Press, 1999).

2. J. Barzun, *Science: The Glorious Entertainment* (New York: Harper and Row, 1964), 116–117.

3. E. A. Hooton, *Apes, Men, and Morons* (New York: Macmillan, 1937), 217–218.

4. J. P. Rushton, *Race, Evolution, and Behavior: A Life-History Approach* (New Brunswick, NJ: Transaction, 1995); L. Lieberman, How Caucasoids got such big crania and why they shrank: From Morton to Rushton, *Current Anthropology* 42 (2001): 63–85; J. Graves, The misuse of life history theory: J. P. Rushton and the pseudoscience of racial hierarchy, in *Race and Intelligence: Separating Science from Myth*, ed. J. Fish, 57–94 (Mahwah, NJ: Lawrence Erlbaum, 2002).

5. W. Saletan, Created equal: Liberal creationism (2007), www.slate.com/id/2178122/entry/2178123 (accessed November 2007); B. G. Charlton, First a hero of science and now a martyr to science: The James Watson affair—political correctness crushes free scientific communication, *Medical Hypotheses* 70 (2008): 1077–1080; J. Malloy, James Watson tells the inconvenient truth: Faces the conse-

quences, *Medical Hypotheses* 70 (2008): 1081–1091; J. P. Rushton and A. R. Jensen, James Watson's most inconvenient truth: Race realism and the moralistic fallacy, *Medical Hypotheses* 71 (2008): 629–640.

6. M. Konner, Seeking universals, *Nature* 415 (2002): 121.

7. S. Pinker, *The Language Instinct* (New York: Harper Perennial, 1995); H. E. Fisher, *The Anatomy of Love* (New York: W. W. Norton, 1992); D. Buss, *The Evolution of Desire* (New York: Basic Books, 1994); R. Thornhill and C. Palmer, *A Natural History of Rape: Biological Bases of Sexual Coercion* (Cambridge, MA: MIT Press, 2000).

8. S. Pinker, *The Blank Slate: The Modern Denial of Human Nature* (New York: Viking Penguin, 2002).

9. F. Boas, New evidence in regard to the instability of human types, *Proceedings of the National Academy of Sciences USA* 2 (1916): 713–718; B. Kaplan, Environment and human plasticity, *American Anthropologist* 56 (1954): 780–800; C. Gravlee, H. Bernard, and W. Leonard, Heredity, environment, and cranial form: A reanalysis of Boas's immigrant data, *American Anthropologist* 105 (2003): 125–138; P. Bateson, D. Barker, T. Clutton-Brock, D. Deb, B. D'Udine, R. Foley, P. Gluckman, K. Godfrey, T. Kirkwood, and M. Lahr, Developmental plasticity and human health, *Nature* 430 (2004): 419–421.

10. M. Strathern, No nature, no culture: The Hagen case, in *Nature, Culture and Gender*, ed. C. MacCormack and M. Strathern, 174–222 (New York: Cambridge University Press, 1980); A. Goodman, D. Heath, and M. Lindee, *Genetic Nature/Culture: Anthropology and Science beyond the Two-Culture Divide* (Berkeley: University of California Press, 2003); S. Jasanoff, ed., *States of Knowledge: The Co-production of Science and Social Order* (New York: Routledge, 2004); D. Haraway, *Simians, Cyborgs, and Women: The Re-invention of Nature* (New York: Routledge, 1991); G. Downey, J. Dumit, and S. Williams, Cyborg anthropology, *Cultural Anthropology* 10 (1995): 264–269.

11. J. P. Jackson Jr., *Science for Segregation* (New York: New York University Press, 2005).

12. R. Bribiescas, *Men* (Cambridge, MA: Harvard University Press, 2006), 12. The ellipsis is standing in for references to Stephen Jay Gould and me, by which I am immensely flattered, although neither of us has ever remotely suggested discounting evolutionary theory for its misuse.

13. Barnes, from a panel discussion on October 26, 2007, at a conference in London titled "Genomics and Society: Today's Answers, Tomorrow's Questions."

14. R. N. Proctor, *Value-Free Science? Purity and Power in Modern Knowledge* (Cambridge, MA: Harvard University Press, 1991); H. Collins and T. Pinch, *The Golem: What Everyone Should Know about Science* (New York: Cambridge University Press, 1993).

15. G. Stone, Both sides now: Fallacies in the genetic-modification wars,

implications for developing countries, and anthropological perspectives, *Current Anthropology* 43 (2002): 611–630.

16. T. Luhrmann, *Of Two Minds: The Growing Disorder in American Psychiatry* (New York: Knopf, 2000); J. Biehl, Life of the mind: The interface of psychopharmaceuticals, domestic economies, and social abandonment, *American Ethnologist* 31 (2004): 475–496.

17. C. Hayden, *When Nature Goes Public: The Making and Unmaking of Bioprospecting in Mexico* (Princeton, NJ: Princeton University Press, 2003); C. Lowe, Making the monkey: How the Togean macaque went from "new form" to "endemic species" in Indonesians' conservation biology, *Cultural Anthropology* 19 (2004): 491–516; P. West, *Conservation Is Our Government Now: The Politics of Ecology in Papua New Guinea* (Durham, NC: Duke University Press, 2006).

18. S. Helmreich, *Silicon Second Nature: Culturing Artificial Life in a Digital World* (Berkeley: University of California Press, 2000).

19. M. Strathern, *Reproducing the Future: Anthropology, Kinship, and the New Reproductive Technologies* (New York: Routledge, 1992); S. Franklin and Helena Ragoné, eds., *Reproducing Reproduction: Kinship, Power, and Technological Innovation* (Philadelphia: University of Pennsylvania Press, 1998); C. Thompson, *Making Parents: The Ontological Choreography of Reproductive Technologies, inside Technology* (Cambridge, MA: MIT Press, 2005).

20. K. Fortun, *Advocacy after Bhopal: Environmentalism, Disaster, New Global Orders* (Chicago: University of Chicago Press, 2001); A. Petryna, *Life Exposed: Biological Citizens after Chernobyl* (Princeton, NJ: Princeton University Press, 2002).

21. F. Bacon, *Meditationes Sacrae. De Hæresibus* (1597).

Index

Text:	10/14 Palatino
Display:	Palatino, Akzidenz Grotesk
Compositor:	BookMatters, Berkeley
Indexer:	Andrew Joron
Printer and binder:	Maple-Vail Book Manfacturing Group